Vampirology

The Science of Horror's Most Famous Fiend

"I simply couldn't get enough of Kathryn Harkup's fascinating, sparkling and erudite account of the history of the vampire. Using actual accounts she demonstrates how the vampire myth spread like a contagion – gossip and rumour mixed with a very modern taste for sensation. How every country and region has its own particularly variety of the blood-sucker, from the *upior* and the *vrykolakas* to the Romanian *strigoi*. And how this area of the world became a melting pot for the superstitions that would give birth to Dracula. She takes us through the endlessly shifting criteria - the Rules of the Beast, as it were – which govern the existence of the vampire: shape-shifting, sunlight and the primary importance of blood itself. She also examines in detail the physical processes of decay and how their misinterpretation could lead the credulous to believe their dear-departed were not so departed after all...

As compulsively readable as a bloated undead feasting on a fresh corpse, this is an absolute must for all children of the night out there.

The blood is the life!"

Mark Gatiss, writer, actor, director and co-creator of BBC's Dracula

Vampirology
The Science of Horror's Most Famous Fiend

By

Kathryn Harkup
UK
harkup.co.uk

ROYAL SOCIETY OF **CHEMISTRY**

Print ISBN: 978-1-83916-157-5
EPUB ISBN: 978-1-83916-412-5

A catalogue record for this book is available from the British Library

The Royal Society of Chemistry is a charity, registered in England and Wales, Number 207890, and a company incorporated in England by Royal Charter (Registered No. RC000524), registered office: Burlington House, Piccadilly, London W1J 0BA, UK, Telephone: +44 (0) 20 7437 8656.

Visit our website at www.rsc.org/books

Printed in the United Kingdom by CPI Group (UK) Ltd, Croydon, CR0 4YY, UK

Acknowledgements

I've been thinking about writing this book for years. But there is a big difference between thinking about a book, or talking about writing a book, and actually doing it. Thank you to the Royal Society of Chemistry, and particularly Drew Gwilliams, for taking it on and for encouraging me to knuckle down and get writing. Drew, and Alethea Doran both gave fantastic feedback and tremendous support during the process.

This book has covered a wide range of different scientific fields and topics. Several people have brought their expertise to bear to check on what I have written. Helen Pilcher dragged my basic and outdated understanding of genetics closer to the incredible modern day achievements in this field. David Jesson was very patient and very helpful when it came to details about wooden stakes. Claire Benson didn't bat an eyelid when I asked her to check my theories about vampire flammability. All of them offered many helpful ideas and suggestions that would never have occurred to me. They have spotted errors and made big improvements to the book as a whole. My heartfelt thanks to all of them.

Many others have scrutinised my scribblings for errors in the general science, history, spelling, punctuation, readability and comprehension. Thanks to all of them and their feedback. The book is in a much better state than it was when they uncomplainingly (at least not to me) read through much earlier,

Vampirology: The Science of Horror's Most Famous Fiend
By Kathryn Harkup

Published by the Royal Society of Chemistry, www.rsc.org

and worse, versions. My appreciation goes to Paul Carpenter, David and Sharon Harkup, David Nevin, Richard Stutely and Mark Whiting. Bill Backhouse and Matt May have also been there to encourage and support me throughout – cheers, you've made the whole process much easier. Special thanks must go to my parents, who unfailingly read everything I ask them to, no matter how badly written it is or uninterested they are in the subject, and still give helpful feedback and unwavering support. Any remaining errors, bad phrasing or incomprehensibility are all down to me.

Dedication

To David and Sharon

Vampirology: The Science of Horror's Most Famous Fiend
By Kathryn Harkup

Published by the Royal Society of Chemistry, www.rsc.org

Contents

Vampirology: The Science of Horror's Most Famous Fiend
By Kathryn Harkup

Published by the Royal Society of Chemistry, www.rsc.org

Prologue

'THERE ARE SUCH THINGS!'
Hamilton Deane as Abraham Van Helsing, *Dracula*

At midnight, one cold January in Marotinu de Sus in Romania, six men opened the grave of Petre Toma and cut out his heart. Something was attacking the people in the village and they believed that something was Petre. He had become one of the undead, and his heart had to be burned to stop him from feeding on the living.

Petre Toma was 76 when he died of cancer. Naturally, his family celebrated his life and mourned his death. The former teacher was laid to rest and the family returned to their lives. But not long after he was buried, Toma's niece complained that her uncle was visiting her in the middle of the night and feeding from her heart, that he had become a *strigoi* (a Romanian word for one who has risen from the grave). She complained he had left her without 'any more blood'. She was very sick. Soon more family members became ill and witnesses confirmed they had seen Petre leaving the house before sunrise.

Toma's brother-in-law, Marinescu, decided he had to act to save his family. Under cover of darkness, Petre Toma's coffin was opened and his corpse inspected for signs of vampirism. His arms had moved and there was dried blood at his mouth. The condition of his remains confirmed their suspicions. Petre had become a *strigoi*.

Vampirology: The Science of Horror's Most Famous Fiend
By Kathryn Harkup

Published by the Royal Society of Chemistry, www.rsc.org

The following night, Marinescu and five other men disinterred Petre's body and cut open the chest with a scythe. The body was full of blood. When the heart was removed, Petre 'sighed'. The heart was impaled on the spike of a pitchfork and held aloft as the men marched out of the village. They stopped at a crossroads where Marinescu's wife and daughter-in-law were waiting for them with a brazier ready to roast the heart. Glowing coals were stuffed into the ventricles of the heart and held up to the night sky. Sparks and flakes of burning flesh flew out in every direction. Every last flake and speck of ash was carefully gathered up and taken to the niece's house. There, the ashes were ground into a fine powder, mixed with water and given to the niece to drink. The following morning she said she felt better and she was declared cured.

This is not fiction. These events were reported in 2004. Those involved were arrested and there are witness statements, autopsy reports and court records to confirm it.

There are dozens of stories like Petre Toma's stretching back over centuries. Many incidents are well documented in reports written by witnesses of unimpeachable character. Tales of the dead returning from their grave and preying on the living, leaving them ill, weakened, sometimes fatally, are not confined to the fiction section of the library and sensational dramatisations on cinema screens. There must be more to this than superstition, mob hysteria and gullibility. What if there really are creatures that feed on the living to sustain themselves in death?

This book is not an exhaustive exploration of vampires and vampire theories, but it will look at all of the classic vampire characteristics and how they might be explained scientifically. Many famous, or infamous, vampires from folklore and fiction will feature along the way, but it is not the intention to determine if these vampires did exist or still do. It is about finding out if vampires *could* exist.

CHAPTER 1

Evolution

> 'Throughout the whole vast shadowy world of ghosts and demons there is no figure so terrible, no figure so dreaded and abhorred, yet dight with such fearful fascination, as the vampire'.
>
> Augustus Montague Summers, *The Vampire, His Kith and Kin*

The mysterious Lord Ruthven certainly made an impression when he appeared at parties. Was it his beauty that fascinated everyone? Or his dead, grey eyes and the deathly hue of his face? Perhaps it was his inscrutable expression and apparent indifference to the company around him that made winning his attentions so desirable. He was certainly like no other creature that had been seen before.

Ruthven was the first modern vampire. He was conceived in July 1816 on the shores of Lake Geneva and made his entrance into London's party scene three years later, on 1 April 1819 (no joke). This sinister central character of the short story, 'The Vampyre', was the creation of John William Polidori, doctor and travelling companion to Lord Byron. He launched the pale aristocratic vampire into the world. It was a monumental moment in literature.

The word 'vampyre' must have conjured up a very strong image in the mind's eye of Polidori's readers. The modern spelling, 'vampire', has changed a little but the mental image it produces has changed a lot. Most people today will see the word

Vampirology: The Science of Horror's Most Famous Fiend
By Kathryn Harkup

Published by the Royal Society of Chemistry, www.rsc.org

'vampire' and think of a tall, pale and handsome figure with sharp teeth and a hankering for blood. But this image is largely the product of mass media; that is, books, theatre, film and TV. All vampire incarnations in the modern mould can be traced back to Polidori's Lord Ruthven. But, those who first read 'The Vampyre' would have been surprised at Ruthven's appearance because, before then, vampires had been very, very different.

Vampires, or vampire-like creatures, have stalked the living throughout recorded human history and in almost every corner of the world. They have appeared under a variety of different names and guises but they have all had some attribute or quality that means they are often included in discussions about vampires.

Over the centuries, these frightening figures have adapted to their environment, just as every creature has. In order to survive the passage of time and changes in culture, the vampire has evolved. The creatures that terrorised ancient civilisations look a lot less threatening today. The vampires of eighteenth-century Europe would not be as successful in the modern world because these stinking, bloated, red-faced beings would be easily spotted and eliminated before they could pose any real threat. Modern vampires have adapted to the modern world by appearing more like us, or like more desirable versions of ourselves, so they can integrate into society and attract their victims.

Still greater variety in vampires can be found around the globe. They all have different names and traits. They can be living or dead, human or demonic, animal or vegetable.[†] They don't all suck blood, and they don't all have fangs,[‡] but they all have some quality that makes them vampiric. To explore the details and characteristics of all these creatures would be an enormous task and well beyond the scope of this book, so we need to narrow things down a little before we go any further.

Definitions are important in science. The subject under scrutiny has to be clearly identified and the boundaries of the investigation carefully marked. So, for our purposes, we will look

---

[†]In an account of his travels in Serbia in the early twentieth century, Tatomir Vukanovic wrote about Muslim Gypsies and their belief in vampiric watermelons – fruit that had been left alone too long and came to pester the living, though they were more of an annoyance than a threat. Spots of blood were said to appear on the melon when it had turned into its vampiric form.

[‡]The Filipino *aswang* has a hollow tongue, and there is a Vietnamese vampire that uses an antennae protruding from its nose to suck blood.

at nocturnal creatures that prey on the living, causing illness and death, predominantly by sucking their victim's blood. Such creatures have become commonly known as vampires, but the origins of this word are as mysterious as the origins of the creatures it refers to. There have been four major schools of thought concerning the etymology of 'vampire'. The Turkish, Greek, Hebrew or Hungarian languages have all been thought to contain the roots of the word. Today, there is general consensus that 'vampire' derives from the Slavic word *bamiiup*. However, regardless of where it originated, the word 'vampire' was introduced into French, German and English literature in the early eighteenth century to describe cases of vampirism in Poland, Russia and Macedonia. Though many parts of the world have their own names for the blood-sucking undead, the word 'vampire' stuck and spread.

Even using a simple definition, there are a huge variety of creatures that fall into the vampire category. Very broadly speaking, there have been three main types, or three ages of vampirism: ancient, folkloric and modern. This chapter is an introduction to these three main eras of vampire and some of the individual species they contained. Looking at it another way, it is a potted history of vampire evolution. Here you will find an overview of some of the blood-sucking undead in their many and gory forms, who we will meet again later in the book. We will see how these creatures have changed and adapted to our society to become the fanged and caped creatures that feature in films, books, cartoons, cereal packets and every children's Halloween party.

Let's begin at the beginning.

1.1 ANCIENT VAMPIRES (3760 BC–1100 AD)

According to some legends, vampires have been with us since the start. According to Jewish mythology, there was Lilith, said to be the first wife of Adam, who was a blood-sucking demon with a particular hatred of children. Lilith's name may have its origins in the Akkadian language of ancient Assyria and Babylonia, where *lilītu* meant spirit.

The Assyrians and Babylonians seem to have had a lot on their plate demon-wise, but one creature with characteristics we would associate with vampires was the *edimmu*. These were the spirits of those who had received an improper burial or died a violent death. They were pale and thin, and their lips were covered in scabs. They stalked ancient Mesopotamia, travelling at night and transforming themselves into dust or mist. *Edimmu* were particularly hostile towards the living. They were known to drink the blood of the dying and spread disease in their wake.

The Assyrian and Babylonian empires co-existed between roughly 2000 BC and 600 BC. Through much of this same period, ancient Egypt was at the height of its powers and prosperity. All three of these ancient civilisations left their marks in the form of great decorative art and buildings, as well as extensive written records, but the ancient Egyptian culture has continued to fascinate the wider world far more than that of other contemporary Middle Eastern empires.

The combination of mysticism, an apparent preoccupation with death and the underworld, alongside very elaborate funeral rites, has led many to expect ancient Egypt to be the source of vampire myths. It has certainly been used in several fictional vampire works as the origin of vampires. However, searches of ancient Egypt's extensive written records have failed to unearth a single vampire-like creature. It may be that their funeral practices were too good. Their lengthy embalming processes and attention to detail meant their dead really were dead and unable to wander about.

As the power of ancient Egypt declined, the region came increasingly under Greek influence and was conquered by Alexander the Great in 332 BC. Ancient Egyptian culture was not erased under Greek control, but the Greeks brought with them their own traditions, some of which merged with the existing Egyptian mythology. The ancient Greeks feared the *Lamiae*: snake-like creatures with the head and torso of a woman. They were said to devour children or drink their blood until they died.[§]

Lamiae were just one of the many types of undead to be found in the ancient Greek world. And readers of Homer's *The Odyssey* will know that the ancient Greek undead liked to drink blood. The ancient Greeks also had the *empusa* to contend with. These vampire-like creatures could change shape to disguise

[§]*Lamiae* are often associated with the Jewish figure of Lilith.

themselves as domestic animals or assume the appearance of humans to fool their victims. The most famous tale about these creatures comes from the *Life of Apollonius of Tyana* by Philostratus, where an *empusa* disguised herself as a beautiful woman to trick a handsome young man called Menippus into fattening himself up so that she could devour him. Thankfully, he was saved in the nick of time by the intervention of Apollonius.¶

The Romans were not left out either. They had ghosts that sucked blood and induced nightmares. Ovid described the *strix*, a ravenous bird or witch that flew at night devouring bodies and sucking the blood of children. Titinius, one of the conspirators against Caesar, noted that a *strix* attack could be prevented by placing garlic on the threatened infant.

Such mythical creatures were not confined to the southern regions of Europe and the Middle East. From Ireland comes the tale of Abhartach, the Dwarf King of Ireland who, according to Irish folklore, ruled over his people some time in the fifth century AD. It is one of the oldest recorded vampire tales and, though it had been circulating orally for centuries, it was only formally written down in 1857 by Patrick Weston Joyce in his *The Origin and History of Irish Names of Places*.

According to the tale, Abhartach was a Celtic warlord and powerful wizard. Those under his harsh rule were keen to get rid of him but feared his dark powers. Rather than tackle the man themselves they hired another warlord, Cathain to do the dirty work. Sure enough, Cathain killed Abhartach and buried him standing up as tradition and etiquette dictated for an Irish warlord. But Abhartach was back the next day demanding a bowl filled with the blood of his former subjects to sustain him. He had become one of the undead, a *neamh-marbh*, or a *dearg-due* ('red blood sucker').

Now the locals had an even bigger problem. Abhartach was still terrorising them, but he could no longer be killed – that had already been done. However, he could be prevented from returning by slaying him with a sword made of yew,‖ burying him upside down and with a huge stone slab on top of the grave

¶It was this story that inspired Keats's vampiric poem *Lamia*.

‖Yew trees have long been associated with the dead, and all parts of the tree, except for the flesh of the red berries, are toxic (see Chapter 12).

to make sure he stayed put. His grave is apparently still there – with its stone slab keeping a lid on Abhartach's final resting place – in the middle of a wheat field. During an attempt to clear the site and remove the slab in 1997, there were several minor incidents and accidents, and the attempt was abandoned. One of these accidents resulted in blood soaking into the ground above the alleged grave. Many now avoid the place.

As well as fending off the *neamh-marbh* and *dearg-due*, the Irish also had blood-sucking fairies and banshees, female spirits whose shrieks heralded a death in the family. These blood-sucking fairies also troubled the Scots, who simultaneously feared the *Leanan Sídhe*, a beautiful fairy mistress who lured men to remote locations to strangle them and then either eat them or drink their blood.

These are not isolated examples: other vampiric creatures can be found in the early myths and legends of many other parts of Europe. Many of these preyed on children – they may have been invoked as a way of dealing with the sudden and inexplicable death of a child at times when infant mortality would have been high. Different entities may have been seen as explanations for the sudden deaths of adults when there was little to no understanding of disease.

As can be seen in many of these examples, the line between mythological creature, or demon, and human during this period in history was extremely blurred. The vampire-like creatures of pre-Christian traditions had a very separate status to the average person. They were often associated with gods – for example, the *empusa* are commanded by the goddess Hecate – and in those days there was a huge reliance on magic to explain unusual or unexpected phenomena. As time went on, societies and their beliefs changed and so did vampires. The folklore vampires that spread terror in later centuries were much more human-like and considerably better integrated into everyday society – albeit on the edges of those societies.

1.2 FOLKLORE VAMPIRES (1720–1816)

The folklore vampire came to prominence in central Europe in the early eighteenth century. These later creatures would be barely recognisable to their ancient predecessors, but they are much closer to the image of the vampire we have today. As Paul Barber puts it in *Vampires, Burial and Death:* 'a typical vampire of folklore ... is a plump Slavic fellow with long fingernails and a stubbly beard, his mouth and left eye open, his face ruddy and swollen. He wears informal attire – in fact, a linen shroud – and he looks for all the world like a dishevelled peasant.'

Several incarnations on from their early forebears, this breed of blood-suckers had adapted particularly well to their environment. The result was a vampire population explosion. They were seemingly everywhere, and reports of vampire epidemics appeared in books and newspapers, fuelling an obsession that has scarcely abated in the three centuries since.

In the eighteenth century in particular, Europe had a huge vampire problem. Villages across the continent were devastated by revenants that tormented and killed members of their community. These vampires were not supernatural spirits; they were far more terrifying, and all because of their ordinariness. They were the undead, former members of the community who were familiar, yet transformed into swollen, stinking, blood-sucking murderers.

It all started in 1731 when a report was published of an investigation into a vampire that had allegedly caused the deaths of more than a dozen people in a village in Serbia. The report became a bestseller at the Leipzig book fair. Translations were made and articles were written.

With so much interest generated in the vampire phenomenon, the hunt was on for more cases, and the searchers didn't have far to look; another prominent case emerged from Serbia. A few years earlier, in 1725, a man had returned from the dead and claimed the lives of ten of his fellow villagers, including his son.** And the more people searched, the more vampires they found. Villages across Central and Eastern Europe saw large numbers of people die suddenly and apparently inexplicably, until a vampire was found in their midst. Vampire epidemics were reported in Istria (1672), East Prussia (1710, 1721 and 1750), Hungary (1725–1730), Silesia (1755), Wallachia (1756) and Russia (1772).

Panic spread and stories multiplied. By 1746, there was sufficient material for Reverend Dom Antoine Augustine Calmet, a French monk and religious scholar, to carry out a thorough

---

**Both these cases will be discussed further in the following chapters.

investigation into the undead in all their forms and publish his findings in *Dissertations on the Apparitions of Angels, of Demons and of Spirits, and on Revenants or Vampires of Hungary, of Bohemia, of Moravia and of Silesia.* It was the first such investigation, but others soon followed.[††]

The church worried that theologians should not be spending time investigating such matters. Calmet's treatise in particular came under censure for not condemning vampirism as outright superstition and uncatholic nonsense. The result was that Calmet's treatise became a bestseller. It was translated into several European languages and fuelled further interest in the undead.

This was the age of Enlightenment, and natural philosophers,[‡‡] as well as theologians, applied their respective expertise to the vampire phenomenon. It was the era when Europe was moving away from a God-centred explanation for all phenomena, to a scientific, evidence-based theory of the universe and all things in it. Treatises started to appear in the eighteenth century applying a logical and rational approach to all manner of things including the undead.

Causes of vampirism were identified in different regions: those who had been buried without proper ceremony; those who had lived an immoral life, or people with the 'wrong' coloured eyes or hair, were all thought to be potential vampires. When several members of a community died suddenly, often complaining of nightmares and visitations from the recently deceased that resulted in difficulty breathing and terrible fatigue, vampirism was suspected.

Graves were investigated for clues: signs of disturbed soil, or graves that animals refused to pass over were suspected of housing vampires. Sometimes, it was said that the vampire could be heard from above the grave, chewing or snuffling like a pig while it fed on its shroud, its own fingers or neighbouring bodies. Suspicious graves were opened and the corpses they contained examined.

Dead bodies were expected to stay still and decay at a reasonable rate. Almost anything out of the ordinary could be taken as a sign of vampirism. Stories of bodies found in their graves,

---

[††]More on his work can be found in Chapter 2.

[‡‡]The term 'scientist' wasn't invented until 1839, almost as a joke. Before then, anyone carrying out scientific pursuits would have called themselves 'natural philosophers'.

in contorted positions, with scratches on the coffin lid, or with their shroud ripped, fuelled rumours of premature burial and the restless undead. Corpses that appeared to have moved in their grave; hadn't decayed; were bloated with blood, or showed signs of blood around the mouth, were considered vampires and appropriate action was taken. Stakes were driven through the chest, heads were cut off, hearts cut out and bodies burnt.

Not all vampires were the same across Europe at this time. There were many variations according to location, language and tradition, and all were carefully catalogued by the vampire hunters of the era. In Poland, the *upior* troubled the population between noon and midnight with its forked tongue and insatiable thirst for blood. Bulgaria had the *ubour* to deal with, an undead creature distinguished by its single nostril, and a barb or sting at the end of its tongue. However, it would only drink blood if food was scarce.

In Greece, Moldavia and the Aegean region, the name given to the dead that pestered the living was *vrykolakas* (or *vroukalakas* or *brucolocas*). Though they did not necessarily drink blood, they could sneak up on sleeping individuals and crush them. Anyone killed by a *vrykolakas* would also become one in death. One Greek island, Santorini, became known as the most vampire-infested place in the whole world. So great was the Santorini vampire problem that the locals became experts in their detection and destruction. Other Greek islanders would bring their troublesome vampires to Santorini for expert disposal.[§§]

But one place in Europe was particularly notable for its mix of cultures, languages and traditions. It also had a correspondingly diverse vampire population. This was the combined regions of Wallachia, Transylvania and Moldavia that are now modern-day Romania. As well as Wallachians, Transylvanians, Moldovians and Romanians, this was the meeting point of people from Germany and Hungary, many of whom settled there. There were also Roma and traders from Russia and Western Europe, and the lands formed the frontier to the Ottoman empire in the south. These mountainous and deeply forested regions isolated communities that then developed their own traditions and ways

[§§]In Greece, taking vampires to Santorini has a similar meaning to 'taking coals to Newcastle'.

of doing things. They also acted as physical barriers to those wishing to explore. Transylvania literally translates as 'the land beyond the forest'.

The many different living people of these regions were troubled by an almost equally diverse group of dead people. *Strigoi* was the Romanian word for the walking dead (the name had been taken from the Roman *striges* or *strix*). These creatures preyed on the living, particularly their families or people they knew, by drinking their blood or other bodily fluids, leaving them weak and sickly. They spread disease wherever they went. They were easy to spot, even by those who hadn't known the *strigoi* in life, with reddish hair that had grown on them in the grave and a florid complexion. They moved slowly, and their resting places were marked by a blue flame burning above their grave. The *strigoi* had to be back in their graves by morning, where they could be found, uncorrupted and slightly bloated from their feeding.

There were also the *moroii*, *moruni*, *nosferatu* and *priculics,* each with their own characteristics and threats to the living. They also each had different ways of being destroyed and different measures had to be taken to protect oneself from each variety of vampire. The *moruni* was native to Wallachia but similar to the *strigoi.* It could transform itself into other animals including cats, dogs and spiders. It killed its victims by draining their blood and, once dead, the victim would also become a *moruni.*

The *nosferatu*, another Romanian vampire species, along with their lust for blood, lusted after other things and was similar to the incubus and succubus of ancient mythology. Any child born to a woman and a *nosferatu* was destined to be a *moroi*, a living vampire or witch. The *priculics* were also blood drinkers, but since they were characterised by their appearance as a wolf, they are perhaps closer to modern definitions of werewolves.

The majority of the vampire-like creatures of this period were the returning corpses of local people. The victims knew and recognised the vampire from knowing them in life. They wore their normal clothes or, more often, their burial clothes or shroud. They had a ruddy complexion and attacked friends and family first, before moving on to the wider community. Such easy means of identifying vampires, and therefore being able to destroy them, obviously threatened the continued existence of these creatures. These stinking, bloated beings needed to become less easy to detect, but not so inconspicuous that they could not find victims. The answer was to get a make-over.

1.3 MODERN VAMPIRES (1816–PRESENT DAY)

John William Polidori, doctor and travelling companion to Lord Byron, was staying with the poet at Villa Diodatti on the shores of Lake Geneva during the summer of 1816. He and Byron had arrived in the region in May of that year and struck up an instant friendship with the poet Percy Shelley and his lover Mary Wollstonecraft Godwin, who would later become his wife. The Shelley party also included Claire Clairmont, Mary's stepsister and Byron's former lover, who would announce she was pregnant by him later that summer. The two groups spent all their time in each other's company.

It is not surprising, given Byron's reputation as 'mad, bad and dangerous to know' and the speculation over the exact relationships between all the members of the group,[¶¶] that rumours and gossip abounded. However, things were almost certainly more dull in reality than the goings-on that were speculated about. The group took in the local sights during the day, and recited poetry and discussed everything and nothing until late into the night. However, their outdoor activities were severely limited. Though the group would rather have been sailing or exploring their magnificent surroundings, they were trapped indoors by appalling weather. Bored with the incessant rain and 'perpetual darkness', but fascinated by the spectacular lightning storms that crashed over the peaks of the surrounding mountains, they entertained themselves by telling ghost stories and discussing science and the nature of life itself.

With no respite in the weather, and searching around for something to entertain themselves, Byron suggested a competition between the five of them: 'We shall each write a ghost story.' The two established writers of the group, Byron and Shelley, didn't finish their stories. Claire Clairmont's story, if she wrote anything, has been lost. Mary, only nineteen at the time, would produce the classic *Frankenstein*. And the twenty-year-old

[¶¶]Polidori initially thought the two women shared Percy Shelley, but he also knew that Claire Clairmont was Byron's lover.

doctor, John Polidori, wrote 'The Vampyre', a tale that was the start of the vampire craze that reached its peak with *Dracula,* and whose influence continues to this day. What might have been a throwaway comment to try and stave off boredom resulted in the creation of two of horror's most iconic figures – Frankenstein's monster and the vampire.

Polidori's short story was published in the *New Monthly Magazine,* but was originally attributed to Byron. Byron denied authorship as vigorously as Polidori claimed it, but the association with the most famous poet of the day did no harm in terms of publicity. 'The Vampyre' was soon spawning sequels, imitations and adaptations for the stage. Theatrical productions opened across London and Paris and proved so popular that one critic complained, 'There is not a theatre in Paris without its vampire!'

Polidori's vampire was very different to those that had gone before, though elements of his forebears remained. This vampire was an aristocrat: suave, handsome, able to mingle in the best company, and giving off few hints as to his true nature. It was the perfect disguise to infiltrate society and lure victims. The vampire in question, Lord Ruthven, was a thinly disguised Lord Byron, who stalked well-to-do society in search of women he could feed from. The story is told from the point of view of Aubrey, who learned about vampires from the beautiful Ianthe. She was one of Ruthven's victims and helped Aubrey realise that if the proposed marriage between Ruthven and his sister went ahead, she would be another victim. This knowledge, and Aubrey's inability to prevent the wedding, triggered a nervous breakdown. His sister died on her wedding night having been drained of blood, and Lord Ruthven escaped.

Ruthven was pale, attractive and from a very different social class to the vampires of folklore, but he was still one of the undead and fed on blood. He was, however, far more picky about his diet than his predecessors, feeding only occasionally, once a year, and only from beautiful young women. Folkloric vampires would feast nightly on relatives, both male and female, and some varieties made do with cows' and sheep's blood. Polidori had added a sexual motive that is not always seen in folklore. Ruthven also appeared to have the ability to be revived by moonlight – a feature not normally seen in the vampires of folklore.

'The Vampyre' inspired many imitations. It has been cited as 'the most influential horror story of all time' by the academic and expert on popular culture, Sir Christopher Frayling. Such recognition came far too late for Polidori. He was paid only £30 (in retrospect) for his story and, two years after its publication, depressed and in debt, he committed suicide by swallowing cyanide.

Although vampires never really disappeared from the stage or from the pages of novels, it was a few decades before the next big milestone in modern vampires arrived. In 1845, Varney the Vampire was introduced to readers of 'penny dreadfuls', serial stories published in weekly parts, each costing a penny. *Varney the Vampire, or The Feast of Blood*, weighed in at over two hundred chapters and a thousand pages. It was a runaway success. Over two years, James Malcolm Rymer thrilled his readers with the adventures of Lord Varney, the aristocratic vampire of the title. He preyed on vulnerable women and underwent a series of near-misses over a few centuries until he (and the publisher), finally tired of his often repetitive adventures, destroyed himself by jumping into the fiery crater of Vesuvius.

Varney borrowed from Polidori's 'The Vampyre' and expanded on it. Slowly, over the course of the nineteenth century, subtle mutations were being made to the vampire and many of the stereotypical characteristics of modern vampires were being established. The undead Varney was tall and gaunt, and his preferred food was blood; he clearly weakened if he did not get regular access to it. He could be 'killed', and often was, but could rejuvenate himself in the moonlight. His several attempts to marry, all thwarted at the last minute, show he had some qualities that made him attractive (even if it was only his wealth).

After *Varney* followed *Carmilla*, a novella written by the Irish author Sheriden Le Fanu and published in 1872. Le Fanu's tale was of a beautiful female vampire that seduced and nearly killed Laura, a teenager who grew up in a 'picturesque and solitary' castle. The action was moved away from England to the romantic and remote location of Styria, a mountainous region of Austria.

The story revealed that Carmilla was not an attractive though listless young woman, but in fact the undead Mircalla – Countess Karnstein – an aristocrat who had been hanging around the area for a couple of centuries. After wiping out an entire village, she sustained her undead self on the blood of local girls. In the dramatic finale she was found inside her tomb floating in the blood from her victims. Le Fanu expanded on Polidori's and Rymer's pale and attractive vampire, ramped up the sex appeal and introduced other elements, either new or borrowed from folklore, that have become staples of vampire fiction. The vampire must rest in its grave, can change shape, has very sharp, pointed teeth and has a hypnotic power over its victims. He also introduced a vampirologist, Baron Vordenburg.

Many other vampire novels and short stories were written during the nineteenth century but, however influential they have been, none came close to the success of Bram Stoker's *Dracula.* Published in 1897 and never out of print since, the influence of this single novel on popular culture is staggering. Count Dracula set the standard for every vampire that followed. The astonishing influence of this novel has overshadowed the many vampire variants that have been described in other parts of the world. European vampires became the dominant species, and it is for this reason that much of this book is Euro-centric.

1.4 WORLD DOMINATION

Stoker's biographer Harry Ludlam claimed that the inspiration for *Dracula* came from a nightmare brought on by 'a too generous helping of dressed crab at supper one night'. There is a long anecdotal history of Gothic horror stories being inspired by nightmares, but even if the initial idea was given to Stoker as a result of indigestion, there were many other sources that could have fed his imagination.

Abraham 'Bram' Stoker was born in Dublin, Ireland, in 1847. Seeds of supernatural stories may have been planted in his mind at a very young age. He was a sickly child and spent much of his youth bedridden, where his mother, born and raised in county Sligo, told him stories from Ireland's rich folklore tradition. Being confined to his bed also gave him ample time to think and daydream, as he later wrote, 'I was naturally thoughtful, and the leisure of long illness gave opportunity for many thoughts which were fruitful according to their kind in later years.' Whatever spooky stories were bubbling away in his mind from his early life in Ireland, he found other important influences when he went to work for the legendary Victorian actor-manager Sir Henry Irving at the Lyceum Theatre in London.

Irving was demanding, self-obsessed and controlling, but Stoker idolised him. It is all but certain that Stoker's manipulative, dominating character of the Count is based on his employer. It was when he was working for Irving that Stoker wrote on a scrap of paper: 'young man goes out – sees girls one tries – to kiss him not on the lips but throat. Old Count interferes – rage and fury diabolical. This man belongs to me I want him.' The note is dated March 1890 and similar phrases appear in Stoker's notes over the following years.

As well as stories from his childhood, and his difficult boss for inspiration, Stoker had almost eighty years of vampire literature, and a wealth of vampire accounts dating from the eighteenth century, to draw on. And he still did a lot of extraordinarily detailed research on top of that, especially when compared with the work he did in preparation for his other, significantly less successful, novels.

He spent time in Whitby, and took a particular interest in local meteorology and shipwrecks, including the Russian schooner the *Dimetry* that foundered in 1885. This ship would become the *Demeter* that would bring Dracula to England in the novel. He visited London Zoo, collected notes on folklore, the supernatural and medicine, even consulting his brother Sir Thornley Stoker, a highly respected medical writer, surgeon and anatomist. The descriptions of Transylvania and Dracula's castle are also very accurate.

Even the superstitions mentioned in the book were taken from real beliefs held by the rural population of Romania at the time. His detailed information about Romania, its traditions and history, he took from an article by E. Gerard titled *Transylvanian Superstitions*. He also consulted a book she had written, *The Land Beyond the Forest: Facts, Figures and Fancies from Transylvania*, full of recollections of her time in Eastern Europe.[|||] Though Stoker's novel was undoubtedly fiction, the many factual elements he incorporated gave the story a realism and credibility.

[|||] Stoker had originally set *Dracula* in Styria, in keeping with Le Fanu's *Carmilla*, but after discovering E. Gerard's work, he felt the romantic and remote setting of Transylvania was the perfect location for the atmospheric setting of the Count's castle.

Perhaps the most interesting factual inclusion in the novel was the vampire himself. The character of Count Dracula, the most famous vampire of all, was based on the real-life Vlad Dracula, born in 1431 in Transylvania and ruler of the adjacent region of Wallachia. Stoker's physical description of the Count was very similar to the portrait of the real-life ruler and Vlad's biography had huge potential for a vampire author.

What really intrigued Stoker was the name. Vlad III Dracula had inherited his fantastic surname from his father, Vlad II, who used the title Dracul because he was ordained as a member of the Order of the Dragon (*draco* in Latin). This group had sworn an oath to the Holy Roman Emperor to defend Christianity from the threat of the Ottoman Turks. Dracula was simply the diminutive of Dracul, 'Dracul's son' or 'son of the dragon'.

Both Vlad the elder and Vlad the younger used images of the dragon as their personal symbol, on their shield and coins. Images of dragons have long been associated with the devil – think of paintings of St George slaying the dragon that symbolised all things evil. And in the Romanian language *drac* means both dragon and devil. Vlad Dracula's name, and the symbol of the dragon, must have given a formidable first impression at the time, but not necessarily a negative one. Dracula was not associating *himself* with the devil, but with one who destroyed evil.

Nevertheless, Vlad Dracula was a ruthless ruler. To maintain law and order within his realm, and intimidate invading Turks, punishments for even minor infringements of the law were strictly enforced, very severe and carried out in public. Vlad's preferred method of execution was impaling on a stake.*** It was said he would eat picnics under the dead and dying bodies, dipping his bread in the blood that dripped from them. However, he was not averse to using other equally horrible methods of execution as he saw fit.†††

It was an extremely effective policy at the time. For example, an ornate and valuable cup was left by a remote spring so that travellers could stop and refresh their thirst comfortably. It must have been tempting to steal the cup, but knowing the punishment

***Through the buttocks, navel or the heart.

†††Including decapitation, hanging, burning, boiling, burying alive and disfigurement by cutting off ears, noses, limbs and sexual organs. There are more, but you get the idea.

for theft, the cup was left where it was. And it wasn't just individuals who could be terrified into good behaviour. When an invading Turkish army approached Dracula's capital city, Tîrgovişte, they were greeted with the 'field of the impaled' – the bodies of thousands of Turkish prisoners impaled on stakes and displayed for the invaders to see. The sight of the rotting corpses and birds making their nests in the skulls of the dead was too much for them, and the army retreated.

Such acts earned Vlad III another title, Vlad Ţepeş ('impaler'). Outside of Romania, he was vilified as a brutal tyrant, particularly in Germany where the first tracts were written about him, with an obvious political bias. Stories of mass impalements and tallies of his victims numbering in the tens of thousands are grossly exaggerated and inflated. For example, one account of Vlad's terrible behaviour, *The Frightening and Truly Extraordinary Story of a Wicked Blood-Drinking Tyrant Called Prince Dracula*, was illustrated with a woodcut depicting Vlad enjoying a picnic under the dead and dying bodies of his victims impaled on a forest of stakes (see Figure 1.1).

Figure 1.1 A German woodcut from 1499 showing Vlad Dracula dining among his impaled victims.

It is simply very difficult, time consuming and labour- and resource-intensive to stake people in such large numbers. Within his native country, Vlad III is seen as a ruthless but effective leader who successfully defended his lands and people. His image has appeared on banknotes and stamps and statues of the man have been erected around the country.

Today there are two distinct Vlad Draculas; the historical figure popular in Romania, and the image of the vampire that is prevalent everywhere else. Apart from an apparent predilection for staking and occasionally dipping his bread in human blood (which may not actually be true), there is little to link Vlad III with vampires. Staking was not unique to Vlad Dracula as it was a common form of capital punishment in Eastern Europe and among the Turks. His association with vampires has everything to do with Bram Stoker and very little to do with the historical record. It's not clear just how much Stoker knew about Ţepeş's personal history, though it was probably very little beyond the name and reputation as a fearsome ruler and great fighter.

Vlad's brutal rule was brought to an end in 1476, when he was killed by a Turkish assassin. His head was taken to Constantinople and displayed on a spike to show the Ottoman power over the West. His body was said to have been taken and buried in the chapel of an island monastery, Snagor, just in front of the altar.[‡‡‡] But that was not the end of the story.

During an excavation between 1931 and 1932, led by Dinu Rosetti, the stone slab that was said to cover the former ruler's remains was removed. The space under the slab was found to be empty apart from a few scattered animal bones. But Vlad Dracula had not risen from the dead to molest the living. A similar slab near the door to the chapel was also raised to reveal a coffin covered in the decaying threads of what had once been an ornate gold embroidered purple pall. Inside the coffin were the remains of a headless body dressed in silk brocade and what was left of a crown. Rosetti believes these are in fact the remains of Vlad III Dracula – reassuringly reduced to a skeleton.

[‡‡‡]Vampires have often had their heads separated from their bodies in order to destroy them, but in Vlad's case it was just easier to transport a head to Turkey than the whole body.

So, apart from a physical description, the background of a brave fighter and a name, Stoker took little from the real-life Vlad Dracula. The rest of Count Dracula's image in the novel owes more to the tradition of Gothic literature than historical fact. No one, not even his worst enemies, claimed Vlad Ţepeş was a vampire of any sort until 1897 and the publication of *Dracula*. The Romanian tourist industry, however, has made a fortune off the back of the association.

Bram Stoker's extensive research and elaboration of fiction with occasional facts paid off. *Dracula* was an instant success and has been the gold standard of vampire fiction ever since. But Count Dracula also differs from what we might identify as a vampire in the twenty-first century. The Count is by no means handsome and attracts his victims using his hypnotic powers. He wanders about in broad daylight with no ill effects and doesn't always wear a cape. Much of the visual aesthetics of the modern vampire are due to stage and screen adaptations.

The first of these was a film produced in 1922 by the German film director F. W. Murnau. He was inspired by the *Dracula* novel, but not enough to pay for the rights to film it. Instead, he changed the names of the main characters and shifted the action to Brennan in Germany. It wasn't enough, and Stoker's widow, Florence, sued, resulting in the destruction of almost every print of the film. One of the most influential films of the twentieth century was almost lost as a result.

Nosferatu, starring Max Shreck as Count Orlock, a vampire who appears to resemble the creatures of folklore rather than the Count found in the *Dracula* novel. He has large pointed ears and sharp rat-like teeth that he uses to pierce the neck of his victim, Ellen Hutter. She is horrified by Orlock's appearance rather than attracted to him and he has to use his impressive hypnotic powers to overcome her.

In 1924, Hamilton Deane adapted Stoker's novel for the stage, this time with full permission from Stoker's widow. He had to make considerable compromises to fit the novel that sprawls over Transylvania, Whitby and London onto a London stage. He confined the whole drama to London. Dracula now appeared in the evenings in a respectable drawing room and therefore did not look out of place in full evening dress. Deane's biggest innovation was to introduce the vampire's most iconic attribute, the long cape. The high collar shielded the face and the cape itself could hide the actor who wore it, allowing for some dramatic stage special effects. Fellow cast members would hold the cloak while the actor playing the Count fell through a trap door and seemed to 'disappear'.

The 1931 film *Dracula* by Tod Browning and starring Bela Lugosi was in turn based on Deane's stage adaptation. This was the first time millions of people had been exposed to vampires in any form. The success of the film meant vampires entered the mainstream. Lugosi's distinctive looks, the jet black hair, long opera cape and thick Hungarian accent set the standard for the characteristics of the vampire (see Figure 1.2). The image of the vampire, that can trace its slowly evolving behaviour and characteristics over centuries, was no longer confined to Europe. It came to dominate the world. Like an invasive species, European vampires have out-competed other vampire-like creatures from non-European cultures.

Figure 1.2 Bela Lugosi as Dracula. An anonymous photograph from 1931, Universal Studios.

Bela Lugosi, Peter Cushing, Christopher Lee and many other actors owe their careers to Dracula and vampires.[§§§] So iconic is Lugosi's portrayal that it was scarcely changed when he became the Sesame Street character that helps kids learn to count. The cape, accent, bats and hair style are all the same. The only omission is the blood-sucking, but it is aimed at kids after all.

The European vampire may have become all-pervasive, but it has not stopped evolving. To maintain its position as an icon of horror, vampires were constantly adapted and reinvented over the course of the twentieth century to suit the mood and cultural expectations of the age.

The Hammer Horror films of the 1960s and 1970s kept Lugosi-style vampires on our screens. In the 1990s there was a resurgence of interest in vampires with *Buffy the Vampire Slayer*, and the novels, then film adaptations, based on Anne Rice's character the vampire Lestat. In the twenty-first century there are animé vampires, superhero vampires and vampires living happily alongside humans. All can trace their direct lineage back to Count Dracula and Polidori's vampyre. Vampires are no longer creatures of mythology, or the revivified corpse of a neighbour wandering around, bloated and bedraggled in their grave clothes. They are smartly dressed, mysterious strangers with sex appeal. Suddenly, the idea of having your blood sucked by such a creature has become almost appealing.

But the vampires of the past have not disappeared completely. The association with bats and wolves harks back to the human–animal hybrids of the ancient vampires. There are also traces of folklore vampires to be found in the vampires of recent novels and TV series. Now we have a better idea of what these creatures are, it is time to see how science can be applied to investigate them.

[§§§]Lugosi was so closely associated with the role of Dracula that when he died in 1956 his family had him buried in his cape and full vampire makeup.

CHAPTER 2

Vampirology

> 'Baron Vordenburg devoted himself to the minute and laborious investigation of the marvellously authenticated tradition of Vampirism.'
>
> Sheridan Le Fanu, *Carmilla*

In 2006, Dr Matteo Borrini, principal lecturer at Liverpool John Moores University, unearthed the skeletal remains of a vampire. He was investigating Lazzaretto Nuovo in Venice, a mass grave containing the remains of hundreds of people. They had all died suddenly, hence their hasty and haphazard burial. But among the thousands of fragmented and commingled human bones, one burial was very different to all the others. The skull of the remains labelled ID6 had a brick rammed into her mouth.

Because of the chaotic way in which the bodies had been buried, it was clear that the position of the brick wasn't an accident. The largest objects surrounding the body were all fragments of bone. The brick had been deliberately placed and careful analysis showed it had been forced into the gaping mouth after death, but before the body had decayed to the point of a disarticulated skeleton. Borrini was intrigued and set about scouring the historical records to find explanations. He came across a treatise written in 1679 by Philip Rohr, *De Mastricatione Mortuorum*, or *On the Chewing Dead*, and suddenly it all made sense.

Vampirology: The Science of Horror's Most Famous Fiend
By Kathryn Harkup

Published by the Royal Society of Chemistry, www.rsc.org

Rohr's treatise explored the phenomenon of dead bodies that appeared to chew their shroud, their fingers and even neighbouring corpses in the graveyard or crypt. From the point of view of Rohr and his contemporaries, ID6 would have been seen as a kind of queen vampire, and the source of death and disease in the neighbourhood; surrounded by her victims. The way to stop the vampire, before it got any ideas about chewing on the living, was to tie up the mouth or block it in some way. Clods of earth or bibles were wedged under chins, and pebbles and bricks were shoved in mouths.

Of course, the skeletons in the mass grave at Lazzaretto Nuovo were not victims of a huge vampire epidemic, with the queen vampire finally stopped in her tracks by a brick; they had died of the plague. These burials dated back to the sixteenth and seventeenth centuries, when plague was a terrifying and deadly reality. The pestilence appeared without warning and struck down huge swathes of the population. In an effort to contain the outbreaks, and to manage the huge death tolls they brought with them, mass graves were dug outside quarantine walls. When the disease reappeared in the city, as it often did, the plague pits were reopened. The disordered state of the burials had not been brought on by the panic and predations of a vampire. It was the panic and overwhelming death toll from waves of plague that often meant digging through layers of victims from previous outbreaks to create more space for the dead.

As gravediggers frantically prepared more graves, they may have noticed that one of the bodies in the plague pit was slightly different. Perhaps one had not decayed as quickly as the others. They probably worried that the open mouth, and the hole that appeared to have been chewed in the burial shroud, was evidence that this woman had become one of the undead, and was responsible for spreading death and disease. The simple solution was to stop her chewing; hence the brick.

Borrini wasn't specifically looking for vampires during his archaeological excavation. Nor were the many other archaeologists who have sometimes uncovered skeletons pinned to their graves with stakes, or with their heads removed and skulls placed at the feet or on the chest. The frequency with which these kinds of burials are found is evidence in support of traditions for dealing with the potentially wandering dead, and beliefs in vampire-like creatures, that stretch back thousands of years.

Many other scientists have inadvertently conducted research that links to vampirism – from chemists looking into new drugs derived from vampire-bat saliva, to forensic scientists studying human decomposition to assist criminal investigations – and we will meet many of them later in this book. But there are some who have dedicated significant amounts of their intellectual lives to the investigation of vampires and vampirism. This chapter is about these vampirologists and their approach to vampirology.

2.1 AN ADVANCEMENT OF LEARNING

Humans are creatures that are curious about the world around them and everything it contains, including vampires. Stories of the blood-sucking undead go back to the beginning of recorded human history because humans were interested enough to write about them. But for a long time very few people looked for explanations for their existence beyond religious interpretations of the phenomenon.

As the previous chapter shows, there were many different types of vampires. Ancient vampires were often mythical human–animal hybrids linked to gods. Though the vampires of the eighteenth century were very different to their ancient forebears, they were still explained in religious terms. These undead creatures were very much human, but at the same time not quite human, and for some, the Devil was quite literally in the detail. Satan himself was thought to be using the remains of some unfortunate humans for his own malicious purposes. Those who had not fulfilled their duties to their faith, or had been buried without the full rights due to them, left a gap that the Devil could use to his advantage, inhabiting the corpse and making it wander again to prey on the living. The undead were evil and the way to combat them was with faith, prayer and, in extreme cases, exorcism.

A God-centred universe was the norm, certainly in Europe, until the late seventeenth and early eighteenth centuries. All things were known to be caused and influenced by God's direct involvement in everyone's everyday lives. But the sudden emergence of vampire epidemics around this time coincided with a huge cultural and intellectual shift in the way people thought about the universe and everything in it. The threat from these blood-sucking revenants appeared to be very real and very dangerous. When it came to combating these creatures, it required more than bravery, strength and faith. You needed science.

This dramatic shift in the way the world was viewed started in the early seventeenth century, with people like Sir Isaac Newton. He had shown that the planets orbited the sun because of gravity. God may have been the Creator, the Great Initiator or Prime Mover in these processes, but the day-to-day movement of these celestial bodies followed fixed mathematical rules and could be explained by simple scientific laws. The universe appeared to run like clockwork, though it may have been God who built the watch and kept it wound.

This new mechanical view of the universe spread. Now, instead of a divine figure being responsible for all the actions in the world, from the movement of the planets to an individual's death, natural philosophers looked for other possible explanations. Inspired by a new systematic approach to science and natural philosophy, careful observations were made, mathematical processes applied and new theories produced to explain phenomena. This new approach yielded incredible results. Early successes propelled many others towards investigating anything and everything. The eighteenth century saw such a staggering rate of progress in scientific understanding and philosophy that it became known as the Age of Enlightenment, or the Age of Reason.

In the scientific world, discovery followed discovery at a bewildering pace. Electricity went from being virtually unheard of, through simple experiments with sparks, and progressed to establishing the basic principles of the new science and the invention of the battery. All in the space of a century. In the late seventeenth century, natural philosophers were still trying to turn base metals into gold. By the end of the eighteenth century, however, chemistry had shrugged off its alchemical past and was established as a rigorous and highly respected science.

Huge advances in anatomy were also being made during this period. Centuries of anatomical dogma was turned on its head, as many probed the inner workings of the body through detailed dissection and observation, rather than trusting to ancient texts. Other medical theories, however, lagged behind. Genetics and the means of biological inheritance were completely unknown and scarcely guessed at. Disease and its various causes were not really understood at all. And there was only a tenuous grasp on things like physiology. This was not down to stupidity; people in the seventeenth century and earlier simply didn't know, and they did their best to find explanations for things with the evidence they had before them and the technology they had at their disposal. Although technology and the sophistication of scientific instruments advanced rapidly during the eighteenth century, it was not even close to what is available today. It only makes the scientific achievements of the Enlightenment all the more impressive because they were done with such limited previous understanding and relatively simple instrumentation.

Anatomists, for example, were undoubtedly highly skilled in their work, but they were severely hampered by the conditions they were working in. No one at that time understood the role of bacteria in decay. During a dissection, individual organs or parts of the body could be removed and preserved for later study, but preserving a whole body was impossible. Anatomists raced to complete their dissections, examining or extracting the most vulnerable organs like the brain and intestines first, but were unable to halt the inevitable deterioration of the body.

Even something as apparently simple as determining if someone was dead or in fact in a deep coma was extremely difficult. Thermometers were first introduced into medical practice in the early eighteenth century by Herman Boerhaave, but it took twenty minutes to get an accurate reading. Improvements were made by Sir Thomas Clifford Allbutt to reduce the time to only five minutes, but that wasn't until 1866. Stethoscopes weren't invented until the nineteenth century.

However, the natural philosophers of the eighteenth century were undeterred. Such was the pace of discovery in areas such as chemistry and electricity that it seemed anyone with care, a keen eye and some rational thought could make significant contributions to the human understanding of the world. There was also little distinction between different scientific fields, or even science and any other areas of learning. In the eighteenth century one could be a poet, a physician and an inventor all at the same time. There were job titles and labels, but people were rarely defined or confined by their area, or areas, of interest. Consequently, anything and everything was up for investigation, and the undead were no exception.

Against this background of feverish scientific endeavour came reports of vampire epidemics. What might have been dismissed in the past as fanciful tales were now subjected to thorough investigation, at least by Enlightenment standards. Several outbreaks were particularly well documented and reports were written by witnesses of unimpeachably good character. The first of these reports that emerged was perhaps the most complete of its type and set the standard.

2.2 ARNOLD PAOLE

In late 1731, reports emerged of a vampire epidemic in Medvegia, near Belgrade. In December, an epidemiologist named Glaster, along with Regimental Field Surgeon, Johannes Flückinger, and two assistant medical officers, Isaac Siedel and Johann Friedrich Baumgartner, were sent to investigate. Flückinger wrote a detailed report of their investigation, which was published in 1732. It caused a sensation.

The report opened with the background to events in Medvegia. The local vampire problem was not new; it had started five years earlier when Arnold Paole returned from the dead to feed on his fellow villagers. Paole had been a *heyduk* (or *hayduk*, or foot soldier)[†] but had resigned from his post near Gossowa in Turkish Serbia because of a vampire attack. On his return to Medvegia, he told several people that while in Gossowa he had been bitten by a vampire. He therefore believed he too would become a vampire and so he had taken preventative measures. He had tracked the undead attacker to his resting place and eaten earth from the grave. He had also bathed his wounds with blood from the vampire, to 'cleanse' himself of the taint of vampirism. He was only partially successful.

[†]*Heyduk* originally referred to a brigand or robber, but in this context meant a band of foot soldiers.

When he returned to his native village, he apparently showed no signs of contamination and carried on with life as normal. Then one day, he fell from a hay wagon and broke his neck. He died a few days later. Twenty to thirty days after his death, he was back tormenting the locals, killing four of them. Paole may have escaped becoming a vampire during his life but he couldn't avoid it in death.

In order to stop Paole from killing more people, his body was disinterred. Forty days after his death, Paole's body was found to be perfectly preserved, though there was blood in his eyes and flowing from his ears and nose. His fingernails and toenails had fallen off but fresh ones were seen growing in their place. This was considered conclusive evidence of Paole's transformation into an arch-vampire. So they drove a stake through his chest. He gave a great shriek and blood spurted everywhere. Just to make sure he wasn't coming back, they burned his body to ashes the same day.

But if Paole had become a vampire because he had been bitten by one, what about the four people Paole had killed? To prevent them also becoming vampires, these bodies were also staked and burned. Satisfied that that was the end of the problem, the villagers went back to their everyday lives. However, their considerable efforts had not been enough.

Five years later, many of the villagers were struck down with feelings of weakness and pains in the chest. After two or three days of growing weaker and weaker, seventeen people died, even though they had been in excellent health before their sudden illness. Men and women of all ages were affected. It was believed that those that had died must have eaten vampire-contaminated meat. Paole, in his earlier blood-sucking rampage, had clearly not restricted himself to humans and had fed on local livestock that had then spread the vampire-contagion to the human population. This was the point where the villagers had called for help and it arrived in the form of Flückinger and his team.

After taking down witness statements about the previous vampire outbreak, the investigators turned their attention to the bodies of the victims of the recent outbreak. All seventeen were disinterred and Flückinger and his colleagues carried out autopsies on all of them. His report detailed their findings case by case. For example: 'A woman named Miliza, aged 60, dead after a three month illness and buried for some ninety days, still had much liquid blood in her breasts ... During her dissection, the heyduks present were astonished to see that her body had become much plumper since her death, and were unanimous in affirming, knowing her as they had done since her youth, that in her lifetime she was extremely thin ... She was judged to be a vampire, for she had eaten some mutton previously contaminated by vampires.'

Eight vampires were identified, the rest of the bodies they examined had decomposed, eliminating them from suspicion. The eight suspected vampires had their heads cut off, their bodies burnt and their ashes scattered in the river. The report finished by asserting that everything it contained had occurred just as had been reported. It was signed by Flückinger and four witnesses.

The report became a bestseller at the annual Leipzig fair and as a result detailed, and often embellished, accounts of the vampire epidemic appeared in magazines and periodicals published in Holland, France and England. Flückinger's position of authority and medical knowledge gave the report credence and the detailed information it contained gave plenty of material to speculate over. A number of things could be concluded from the report:

1. Humans could turn into vampires if they were bitten by one (see Chapter 8).
2. Vampires didn't die and decay as humans were expected to. They showed evidence of life because parts of their body continued to grow, they cried out when they were staked and they were full of blood (see Chapter 3).
3. Vampires drank blood, weakening and killing their victims (see Chapter 4).
4. They operated at night (see Chapter 5).
5. You couldn't talk nicely to a vampire or gently persuade them to desist. Extra measures had to be taken to prevent further attacks (see Chapter 11).
6. Stakes and cremation could be used to destroy vampires, but you had to be thorough. Any vampiric beings that slipped through the net could come back to cause problems at a later date (Chapter 12).

At no point, however, did Flückinger offer any opinion or explanation for his observations. He detailed what he had witnessed and the actions that were taken, but offered no theories as to the underlying cause of the unusual things he had witnessed. Others, however, were more than willing to fill in the gaps. Following Flückinger's report, many people applied their minds, and then put pen to paper, to try and explain these episodes of vampirism. Over the period of a few short years following Paole's predations on the people of Medvegia, there was a small avalanche of books, dissertations and treatises speculating on the extraordinary events that had been reported. Many theories were offered to try and explain the phenomenon, from the spiritual to the scientific.

2.3 THE GOLDEN AGE OF VAMPIROLOGY

Vampires and vampirology were suddenly in vogue, and a number of figures emerged as experts in the field. Christopher Rohl was an early eighteenth-century theologian who wrote a *Dissertation on the Bloodsucking Dead* in collaboration with Johann Hertel, which was published in Leipzig in 1732. It was rated highly amongst several other similar works published at the time. In 1733, Johann Heinrich Zopfius, a teacher and vampire expert; and Karl Francis von Dalen, about whom little is known, together wrote *Dissertation on the Vampires*. Another early eighteenth-century writer and vampire expert was John Christian Harenburg, who wrote several works on his favourite subject in the 1730s, most notably *On the Vampires* and *Philosophical and Christian Thinking on the Vampire*.

Various theories were put forward, but the idea that the devil was behind it all, inhabiting corpses and moving them to do his bidding, was still prominent. Others searched for a more natural, or medical, explanation. Some thought the evidence found at the autopsies suggested premature burial. Could the symptoms of alleged vampire attacks be explained by contagious food poisoning? Could it be the effects of opium, or other drugs?

One man, rather than trying to explain the whole series of events, focussed on a specific aspect. Jean Baptiste de Boyer, the Marquis d'Argens, set about investigating the fresh blood reported to be present in the graves of vampires. Writing in 1738 in his *Lettres Juives*, he asserted that 'juices in the dead body by nitreas particles fermenting with them, may furnish a liquor nearly resembling blood.' Those carrying out the investigations into cases of vampirism may have thought they saw fresh blood, but they were mistaken, or had misinterpreted their observations. Other writers questioned if soil conditions could have preserved some corpses better that others, or if these epidemics were caused by disease rather than the undead.

By and large, the writing on vampires up to this point was in the form of dissertations or short pieces included in much longer works that covered a range of topics. It wasn't until 1746 that Reverend Father Dom Augustin Calmet, who we met in the previous chapter, attempted to collate all the known examples of reanimated corpses and wandering spirits and seek explanations for them.

Many of Calmet's critics said he was too gullible – too trusting of the accounts and too willing to accept vampires as a real phenomena – though he made no such claim in the book. He took the evidence as it was presented and asked sensible questions. He offered down-to-earth explanations for many of the characteristics that were attributed to vampirism. For example, the cry given out by the vampire when it was staked was merely gas escaping through the throat as pressure is applied to the chest. Evidence of movement or dead bodies eating could be because people had been buried alive – there were many well-documented cases of people reviving after they were thought to be dead. He also discussed the issue of how, in one context, a well-preserved corpse would be a sign of saintliness, but in another it would be seen as a sign of vampirism. Calmet's answer was that vampires smelled bad, of decay and corruption, but saints did not give off such offensive odours and in fact often smelled of nice things like flowers.

However, he, like many natural philosophers during the Enlightenment, could not answer every question that was posed. Calmet could not explain how vampires could escape and return to their graves without disturbing the soil about the coffin, and admitted as much in his book. His incomplete explanation was seen by some as a failure, but it was very much in the spirit of scientific investigation. At the time, natural philosophers were not afraid to speculate. It was acknowledged that not everyone had the answers to everything and the search for those answers was open to anyone who wished to apply themselves to the problem. Calmet's acceptance of accounts, taken at face value, even though he did not necessarily believe that what witnesses saw was supernatural, was also seen as a failure by the Church authorities and tantamount to a belief in vampires.

In spite of, or perhaps because of, the controversy over his work, Calmet's *Dissertations on the Apparitions of Angels, of Demons and of Spirits, and on Revenants or Vampires of Hungary, of Bohemia, of Moravia and of Silesia* became another vampire bestseller. Published in 1746 in Paris, with an English translation appearing in London not long after, it was still being quoted and discussed over a century later. Calmet, and his predecessors, had established vampires as worthy of closer scrutiny and vampirologists were recognised experts in this field.

2.4 THE NINETEENTH CENTURY

The pace of scientific advances in the eighteenth century had been extraordinary, and continued to accelerate as the century came to a close. New discoveries were made at such an astonishing rate, it was difficult to keep up. By the turn of the nineteenth century, science was being discussed in lecture theatres, coffee shops and well-to-do drawing rooms. Public demonstrations of chemical reactions and electrical phenomena were hugely popular, causing traffic jams in the streets outside London's newly opened Royal Institution. Scientific books and pamphlets were published and the latest scientific equipment could be purchased in the shops. To many it was an exhilarating time of scientific advancement, but for others it was terrifying.

One reaction to the dramatic pace of change was the Romantic movement. When nature appeared to be about to give up her last secrets, artists reacted against this rational approach to everything and looked to the past, and to the natural world and the inexplicable. Romantic poets introduced vampire themes into their works, but took their inspiration from ancient beliefs in mythological creatures – Johann Wolfgang von Goethe (*The Bride of Corinth*, 1797), Robert Southey (*Thalaba the Destroyer*, 1801), Lord Byron (*The Giaour*, 1813), Samuel Taylor Coleridge

(*Christabel*, 1816) and John Keats (*La Belle Dame sans Merci*, 1819 and *Lamia*, 1820). But folklore vampires could not be kept away, and traces of them crept into the characteristics of these poetical vampires. The Romantics may have been reacting against the progress of science, but they were also fascinated by it. And there was nothing they could do to stop or even slow its advance.

As the nineteenth century progressed, the rational approach was reintroduced to vampire stories. The role of scientific investigation and experts in vampirism took an increasingly central role in the vampire stories produced over the course of the century. In 1819, in Polidori's 'The Vampyre', the character Aubrey knew very little about vampires and was unable to stop Lord Ruthven. There was no vampirologist on hand to help and the vampire escaped. On that occasion, the vampire had won but both vampire and vampire hunter evolved in an almost symbiotic predator–prey relationship.

Vampires had evolved over the centuries to outwit those that wished to destroy them. The blood-sucking undead of the nineteenth century were certainly smarter than their folklore predecessors and humans had to learn fast to keep up. The way to defeat these intelligent, calculating and resourceful creatures was to study them. Scientists, or more specifically vampirologists, came to the fore.

2.5 FICTIONAL VAMPIROLOGISTS

Some of the most important fictional works in the vampire genre from the nineteenth century – *Varney, the Vampire*, *Carmilla* and, of course, *Dracula* – included vampirologists or scientific approaches to the unusual events depicted in the novels. The titular vampire in *Varney* is thwarted in his many attempts to marry, and feed, on young women, by the Bannerworth family (see Figure 2.1). After a series of attacks on young Flora Bannerworth, the family, plus Flora's fiancé, Charles Holland, his seafaring uncle, Admiral Bell, and Admiral Bell's mate, Jack, combine forces to investigate. They were perhaps the nineteenth-century ancestors of TV's *Buffy The Vampire Slayer*'s Scooby Gang. They closely examined the wounds to Flora's neck and her health was monitored. Suspicion fell on the mysterious new neighbour. They interviewed the neighbour, observed his behaviour, studied

Figure 2.1 Title page from *Varney the Vampire or the Feast of Blood.*

his habits, researched into vampire lore, opened graves, and tested different means of slaying their foe. The Bannerworths were essentially amateurs learning on the job, but their approach was measured and fairly scientific. Their actions contrasted dramatically with the baying mobs that sought revenge for

Varney's attacks. They were angry, armed and caused no end of havoc, but they never caught the vampire.‡

Le Fanu's *Carmilla*, published nearly thirty years after *Varney*, introduced a level of professionalism into vampirology. Baron Vordenburg was the descendent of the vampire Carmilla's early victims who 'devoted himself to the minute and laborious investigation of the marvellously authenticated tradition of vampirism.' It was thanks to his detailed knowledge that the tomb that housed the vampire was discovered and its inhabitant destroyed.

Baron Vordenburg was a relatively minor character in Le Fanu's story; the main focus was on the vampire and her victim. When Bram Stoker came to write his vampire novel, the emphasis was very different. Despite having his name as the title of the book, Dracula hardly appears in the story (though he was discussed at length), it was those trying to destroy him who took centre stage. The small team of amateur vampire hunters was led by an expert, the most famous vampirologist of them all, Professor Abraham Van Helsing MD, D.Ph., D. Litt., *etc.*

Van Helsing was the real hero of *Dracula*: the scientist who triumphed over the superstitious world of the Count. The Dutch polymath may be the most recognised vampirologist, but his research into all things relating to the undead was merely a sideline to his main career. The professor had to draw on all of his vast knowledge of philosophy, metaphysics and the law, as well as carrying out additional research into folklore traditions and superstitions, in order to defeat Dracula.

The Professor was initially consulted when Lucy Westenra fell ill with a disease that her suitor and doctor, John Seward, could not diagnose. 'She eats well and sleeps well, and enjoys the fresh air; but all the time the roses in her cheeks are fading, and she gets weaker and more languid day by day; at night I hear her gasping for air.' At first glance, it would appear to be a classic case of tuberculosis (TB or 'consumption' as it was known until the twentieth century). The disease was horrifyingly common in the late nineteenth century (see Chapter 8), so diagnosing it would have been straightforward for a doctor of the time.

‡The now well-worn trope of angry villagers wielding torches and pitchforks originated in *Varney, the Vampire*.

But there were other symptoms that could not be explained; marks on Lucy's neck that wouldn't heal; a bloodlessness that wasn't anaemia, and Seward was stumped as to the cause of her declining health. As 'one of the most advanced scientists of his day' and an expert in rare diseases, Van Helsing was the obvious person to bring his knowledge to bear on the problem.

Lucy's extreme blood loss was threatening her life, so Van Helsing used blood transfusions, an up-to-date but high-risk treatment, to bring her out of immediate danger. But the answers to the cause of the blood loss were not to be found in science. Van Helsing consulted 'the teachings and records of the past', as well as experts to confirm his theories about the true reason for Lucy's illness. Stoker even name-dropped a real-life expert in history and culture in his novel, the distinguished Orientalist Ármin (or Arminius) Vámbéry from the University of Budapest. The author had met Vámbéry at dinner a month after he had started working on *Dracula* and much of the discussion that evening had been concerned with the supernatural. In the novel, Vámbéry was described as a world authority on vampires, but in reality he was an expert in the cultures and languages of people living in Eastern Europe and the Middle East.

Over the course of the novel, Stoker had Van Helsing conduct a thorough investigation into the nature of vampires and vampirism. The Professor used careful observation, accounts from witnesses, as well as existing knowledge to form theories about Dracula. He constantly tested his ideas and learned from the results. Though he couldn't save Lucy, his combined knowledge of history, folklore traditions and science saved other victims and enabled his team of vampire hunters to track down the Count.

In creating the character of Van Helsing, Stoker had borrowed elements from the eighteenth-century writers and experts on vampirism like Calmet and the Marquis D'Argens. Their rational approach to understanding seemingly irrational events was blended with the mystical and supernatural portrayal of vampire-like creatures by the Romantic poets. Stoker also took inspiration from the prose writers Rymer and Le Fanu. In *Varney*, the Bannerworth band of vampire hunters had set the trend for small groups of slayers, but Stoker brought them to prominence. Bands of slayers, whose members brought a range of skills, from strength and bravery, to scholarly interests in old texts, as well as scientific knowledge, became a regular feature of vampire films, TV and books throughout the twentieth century. The study of vampires became almost as important as the vampires themselves.

2.6 TWENTIETH-CENTURY VAMPIROLOGY

The early twentieth century saw a throwback to the early days of vampirology in the shape of Augustus Montague Summers. Summers was an unusual character even by vampire standards. He was an English cleric and Gothic scholar who styled himself as the 'Reverend Alphonsus Joseph-Mary Augustus Montague Summers' and wandered around in the elaborate garb of a Catholic priest, although there is no evidence he was ever ordained (his interest in Satanism and the occult, along with rumours of improper relations with young boys probably put paid to that).

Summers wrote several books on occult themes, including witchcraft and werewolves, but he also produced two volumes on vampires, *The Vampire, His Kith and Kin*, and *The Vampire In Europe*. These books were an accumulation of every vampire-related story Summers could find, similar to Calmet's work from two centuries earlier but much more extensive. Summers attempted to explain the vampiric phenomenon described in these accounts, and exploited the most tenuous links to vampires as proof of their existence. For example, when discussing Calmet's doubts over how a vampire could enter and exit its grave without disturbing the soil, Summers found an answer in spiritualism.

Spiritualism was having its heyday in the early twentieth century, séances were all the rage, mediums were contacting the spirit world, manipulating ouija boards and producing ectoplasm. Summers thought the way mediums claimed to use spirit energy to form physical ectoplasm might also apply to vampires entering and exiting their coffins. He suggested vampires could manipulate their physical bodies into a 'subtle' form that could pass through material objects. The vampire's victims provided the energy to enable the transformation.[§]

[§]The scientific credibility of these assertions will be explored in more detail in Chapter 7.

Summers included every scrap of information he could find and he filled the pages with quotes and abstracts in their original German, Greek, French and Latin; perhaps to give the work a scholarly air, or because he was rushing to complete his books and didn't have time to translate them. He even found space for fictional vampire works such as *Varney, the Vampire* and speculated if the author had in fact discovered a real but previously unknown vampire incident. When writing *Varney*, Rymer had clearly taken the trouble to find out about 'real-life' vampire epidemics such as the one mentioned earlier that was caused by Arnold Paole. Occasional references to well-known cases and brief discussions of science made the rambling story a more interesting read, but perhaps mislead the rather gullible Summers when he read it.

Dracula, on the other hand, was a disappointment for Summers because of what he saw as Stoker's use of 'more extravagant details of vampirism which frankly have no place outside the stories told round a winter's hearth.' He felt the author should have 'confined himself to those particulars which are known and accepted, which indeed have been officially certified and definitely proved.'

Exactly what Summers himself believed is more difficult to judge. He appears to have believed in vampires wholeheartedly, but it might also have been part of his eccentric personality and he may have been writing with his tongue firmly in his cheek. His often bizarre compendiums of all things vampiric may not be the most accessible books on the subject, but they have achieved a kind of a cult status, along with Summers himself, something he probably would have relished.

Many other twentieth-century scholars have written about vampires in all their forms, and offered explanations for their existence. There are many folklorists that specialise in the study of vampire legends and tales of vampire outbreaks. Rather than scouring the records and plucking out anything that looks like a vampire, folklorists and cultural historians have written about vampires within the context of the communities that believed in them. There are anthropologists who study burial practices or research into the beliefs and traditions centred around blood-suckers in various cultures.

As we saw earlier in the chapter, archaeologists have uncovered the remains of humans staked, decapitated, with bricks shoved in their mouths and other methods commonly used to prevent the dead rising or feeding on others. Researchers who work on animal behaviour, human diseases, drug therapies or nutrition have come across symptoms and side effects that remind them of the vampires they are familiar with from books, TV or films. The majority of these researchers are not looking specifically for vampires, but have come across instances that match popular conceptions of vampire characteristics. Many of these researchers have looked at a few well-known characteristics of vampires and found rational explanations for them. Some have gone further and researched into the history of vampires and put forward theories as to how vampirism as a whole could be explained by modern science. These researchers might not call themselves vampirologists, but their cumulative contributions to explaining vampire phenomena could be called vampirology.

Vampirology is multidisciplinary. The Venice vampire we met at the start of this chapter is a prime example of how something a bit different, a bit unusual, catches someone's eye and that difference is explained by pulling together many diverse strands of knowledge. The Venice vampire was noticed because she was different from the other burials around her. Hundreds of years ago, those differences were understood in terms of people's expectations of what happened to a person's body when they died, their faith and local traditions. Centuries later, the skull with the brick in the mouth again stood out from the hundreds of other skulls it was surrounded by. The remains of the Venice vampire were uncovered by archaeologists, analysed by scientists and interpreted using historical records and in terms of folklore traditions.

Vampires are clearly different from humans and other creatures. It is these differences that can be examined to understand what they could be and how they might exist in the world. The Venice vampire stood out because she was dead, but not behaving as dead people were expected to behave. What happens to the dead, and the undead, is the subject of the next chapter.

CHAPTER 3

Undead

> 'And many of those who sleep in the dust of the earth shall awake'
>
> Daniel 12:2

Late one night, Peter Plogojowitz came home and demanded food. His terrified wife gave him some and he went away again. As soon as he had left, Peter's wife packed up her things and went to stay with relatives in another village. She had hoped never to see her husband again. You see, she had buried him three days earlier.

Peter's wife was lucky to escape when she did. Two nights later, Peter was back at his former home demanding more food. This time, it was his son who answered the knock on the door, but his request was refused and Peter went away. The following morning, the son was found dead in his bed. Then more people in the village died. They would be taken ill very suddenly with exhaustion and they seemed to have lost a lot of blood. Some claimed they had dreams of Peter visiting them in their sleep and feeding from them. In the space of just eight days, nine people died, and Peter was blamed for them all. A local magistrate decided something had to be done and so he contacted the Imperial Provisor, Gradask District (in Austrian-controlled Serbia), asking for help with the village's vampire problem.

Vampirology: The Science of Horror's Most Famous Fiend
By Kathryn Harkup

Published by the Royal Society of Chemistry, www.rsc.org

The commander of the imperial forces in Gradask made a personal visit to the village along with two other officers and an executioner. He ordered the exhumation of Peter Plogojowitz and all his alleged victims. By this time, Peter's body had been buried for some ten weeks, yet it was found to be in a remarkable state of preservation. Only his nose showed slight signs of decay. His body was flushed and plump, as if he was full of blood. His hair and nails appeared to have grown and his eyes were open. His skin had sloughed off but fresh skin appeared to be growing underneath. The edges of his mouth were smeared with what looked like fresh blood. All clear signs of vampirism in the eyes of those present.

A stake was driven through Peter's chest and blood gushed everywhere. His body was then cremated. His victims were also examined for signs of vampirism. Those thought to have become vampires were also burned. The rest were reburied, but with sprigs of whitethorn, lots of garlic and other herbs, just in case. This was apparently enough, as no further attacks were reported.

This is not the plot of a nineteenth-century Gothic novel; it was reported as fact on 21 July 1725 in the Viennese newspaper *Wienerisches Diarium.* Newspapers have been known to exaggerate on occasion, and the article may have contained embellishments and salacious speculation, but the source of the story was impeccable. The Plogojowitz case, with its many witnesses and an official report, added to the growing number of vampire epidemics. The detailed examination of Peter's corpse appeared to add to the weight of evidence for the existence of the undead. But what was really going on?

3.1 TO BE OR NOT TO BE?

With a few notable exceptions, the majority of vampires in folklore, films and books are dead.[†] It is one of the most important characteristics that distinguish vampires from humans. They have ceased to be and are not simply 'resting'. And though vampires may have shuffled off this mortal coil, they certainly haven't joined the choir invisible.

[†]Ancient vampires might be an exception but the debate over whether demons and gods are living in the same sense as humans is a philosophical question for another book.

Checking for a heartbeat or a warm body might seem like an obvious and easy way to distinguish between the living and the undead. But if your 'vampire' is up and about, do you really want to get close enough to them to feel their pulse or insert a thermometer somewhere? It seems like a very high-risk strategy to me. It's much better to wait until they are back in their coffin resting. A quick look at the date of death and an examination of the body should make it easy to establish if the coffin's inhabitant is really dead or not. Unsurprisingly, this calm, low-risk approach has been the one taken by the majority of those seeking vampires.

In epidemics where a vampire was the suspected cause, such as the case of Peter Plogojowitz and Arnold Paole from the previous chapter, initial fears were raised when the recently deceased were seen wandering about when they should have been quietly rotting in their grave. Locals had the good sense not to approach these figures in person and check for a pulse. But then that would have been unnecessary; they already knew the person was dead. Along with many other folklore vampires, these people were known to their victims during life and had in all probability attended their funeral. Is there any chance they could have made a mistake?

Diagnosing death can be tricky, even today with the help of sophisticated medical equipment. Death is not a light-switch moment, that changes from on to off, it is a process. Different cells die at different rates in the body. Some cells, such as nerve cells, can die in minutes but others, like fibroblasts forming the connective tissues, take days. But we rarely wait for all the everyday processes of life to stop before we declare someone to be dead.

The current medical point of death is when a person's brain has died. Nerves in the body may continue to cause small movements, the heart may still be beating, and if the individual is connected to an artificial respirator, they will still be breathing. However, if tests reveal there are no signals being sent from the brain stem to the rest of the body, then the person has reached a point of no return. They can be declared dead, as there is currently no medical treatment that can fully restore function to those damaged parts of the brain.

In the eighteenth century, when Peter and Arnold were declared dead, things were much more simple. If the heart had stopped beating, they were dead. No pulse, no life. Lack of movement or signs of breathing and a cold body would have confirmed the diagnosis. But there are problems with this straightforward means of determining death. People can be paralysed in a catatonic state or a coma. A pulse can be feeble and breathing shallow, and therefore both can be difficult to detect. People can be cold and alive, and bodies can be warm and dead. Inevitably mistakes have been made.

Stories of people reviving on the way to their own funeral have been recorded since the time of the Romans. The ancient Greeks, aware of the possibility of getting it wrong, used to cut a finger off the dead, just to check. As time and medical knowledge advanced, the difficulties of diagnosing death only became more apparent (see Figure 3.1).

The surge in interest in anatomy in the eighteenth century meant that anatomy theatres were popping up all over the place. One source of material for the professors and students was the scaffold. In Great Britain, the bodies of convicted murderers were left hanging for an hour before they were cut down. The body might have long stopped moving, and the face may have been bloated with blood; the tongue, blackened and swollen, might protrude from the mouth, but they were not necessarily dead. The body, apparently bereft of life, would be taken directly to the anatomy theatre fresh from the place of execution. Professors of anatomy would cut into the chest and

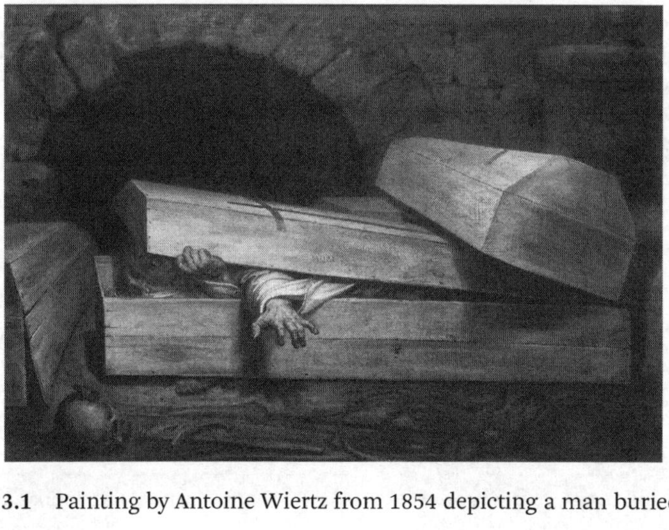

Figure 3.1 Painting by Antoine Wiertz from 1854 depicting a man buried alive.

hold up the dissected heart to gasps of astonishment from the students, who could see it was still beating.

These mistakes, however, would have been made in only a tiny minority of cases. The scientific detail of life and death may not have been understood in the eighteenth century, but the vast majority of people would have understood what death looked like. Unlike today, most people would have died at home, cared for in their last moments by loved ones. Personal experience would have made death a safe diagnosis in the overwhelming majority of cases. And when someone died, the body would have been kept at home while it was prepared for burial by relatives or local women. The days of professional morticians, who remove bodies and prepare them out of sight, were in the future. This would have given many in the eighteenth century, and earlier, a much better insight into the early stages of decay than the majority of people today.

When it came to investigating vampires, the condition of the corpse was particularly important. For example, in the Peter Plogojowitz case, his body was examined long after death, but it did not look the way it was expected to look. The same was true of the nine vampires in the Arnold Paole outbreak (see Chapter 2). It was often stated in these reports that the body had not decayed, but then immediately went on to describe signs of decay, just not the ones that were expected. Any and all of these anomalies could be taken as signs of vampirism. Arnold Paole and his eight victims didn't even show the same pattern of anomalies but all were labelled vampires nonetheless. Any change in the body that was out of the ordinary was taken as evidence that the dead person had become a vampire. Before we explore these *ab*normal observations, we should find out what is normal.

3.2 DEATH IS NOT THE END

The natural way of all things is towards a more disordered state.[‡] Life could therefore be described as the ability to fend off decay. Our bodies are constantly being damaged, broken down and decayed, but we have mechanisms that repair damage, limit decay and encourage new growth. To go against the normal run

[‡]The second law of thermodynamics applies to everything, with no exceptions.

of things, to maintain the body in its highly organised state, requires a constant supply of energy.

Eating food and breathing in oxygen gives us the raw materials that are combined in a series of chemical steps, carefully orchestrated by enzymes, to release carbon dioxide, water and the all-important energy. This process, called respiration, is carried out inside almost every cell of our body, so there is a ready supply on tap for the cell to do its work. It's like every home having its own personal power station so it doesn't have to rely on an external supply. In our homes the majority of our devices are powered by electricity, but in the body it is chemical energy that is used.

The energy released by the respiration process is stored up in chemical bonds. Three phosphate units are added to adenosine to form adenosine triphosphate (ATP). The phosphates can then be broken off from the adenosine one by one to release the stored energy to facilitate chemical reactions and allow molecular machinery to operate smoothly.

When a person dies, they stop eating and breathing and therefore respiration stops and the supply of energy is cut off.[§] Without energy, cells cannot carry out even the most basic functions within the body and the result is decay, a process that can be very roughly summarised in four stages: fresh, bloat, active decay and skeletonisation. This simple categorisation masks a fantastically complex series of events. The process starts almost immediately and proceeds rapidly, driven by chemical processes that are normally balanced or kept in check by the living body.

One of the most basic functions of the body is to expel waste carbon dioxide produced from respiration. Oxygen is drawn into the body and is quickly used up, but when a person is dead, breathing has stopped and so the carbon dioxide produced from the last breath cannot be expelled. Carbon dioxide dissolves in water to form a weak acid (carbonic acid), lowering the pH of the cells where it was produced. The increased acidity weakens the cell membranes, causing them to rupture and spill their contents into the surroundings.

[§]People who have been declared brain dead can be connected to respirators to allow the continued flow of oxygen that keeps organs functioning and viable for transplant.

Tissues in the body begin to be digested from within in a process called autolysis. Enzymes from within the cell, those that break bonds and make chemical changes essential for the construction of new components for the body, are now free of their controlled environment. Rather than carrying out their chemical tasks in well-defined processes within the cell, they can now break bonds and carry out chemical reactions anywhere. The indiscriminate chemical bond breaking and making accelerates the rupture of more cells. The process can begin within minutes of the final breath but remains hidden away inside the body, at least for the time being.

The first visible signs of decay occur around half an hour after death. Without the heart beating to pump blood around the body, gravity slowly pulls the blood down and it begins to pool in the lowest parts. This pooling of blood will initially appear as purplish-red blotches, and the uppermost parts of the body will look pale by comparison. As more blood settles and congeals it becomes fixed in its position and the colour darkens. The upper parts of the body may appear flat or depressed as the fluid that normally plumps up the tissue has drained away to the lower parts and become trapped there. Now, even if the body is moved, the dark purple and pale patches will remain in place. These colour changes are known as livor mortis and continue after the blood has settled and set in its position.

Around the same time as livor mortis becomes obvious, the first signs of rigor mortis can appear. This is another chemical consequence of stopping respiration – diffusion. During life, certain elements and molecules are held in high concentrations at specific sites within the body so they can function normally. For example, a low concentration of calcium inside muscle cells means the cell is relaxed, but a high concentration causes contraction. Respiration provides the energy that powers the molecular pumps controlling the flow of calcium in and out of the cell. Without a supply of energy to keep molecules in their right place, diffusion takes over and chemicals will leach out of their concentrated pockets to spread into an even distribution. And as calcium diffuses into muscle cells, they begin to contract.

Other factors also contribute to the progress of rigor mortis. ATP, as well as being the body's store of energy, carries out many other functions, including allowing muscle fibres to glide over each other smoothly. Whatever reserves of ATP the body has left over from the last breath are quickly used up, and actin and myosin fibres in the muscles will begin to bind to each other, stiffening the muscle. Over the course of a few hours, with no more ATP available and more calcium leaching into cells, the muscles become locked in position and remain so for two to three days.

Rigor mortis will wear off and the limbs become supple again, but not because the muscles have relaxed. After two or three days, the autolysis process that has been quietly working away within the tissues of the body will eventually digest the muscles to the point that they can no longer maintain their rigid position. The two-day mark is something of a tipping point for autolysis where its effects become much more apparent, and not just in the disappearance of rigor mortis.

The process of autolysis is aided and abetted by microbes in the body. Bacteria benefit from the part-digested proteins and other products of the enzymes' unchecked work on tissues in the body. The ready supply of food means their numbers swell and spread throughout the body. When the breakdown of the tissues breaches the lining of the gut, there is an explosion of activity. There are trillions of bacteria living in our digestive system and benefiting from a warm and protected environment and a steady supply of food. In return for giving the bacteria a safe home we benefit from them because they break down foods we would otherwise be unable to digest. These colonies of microbes will continue to metabolise, grow and multiply, as long as conditions are favourable.

When a human being dies, their colonies of gut bacteria do not die along with them. These microbes will continue to survive as long as possible on the body's last meal. When the bacteria run out of the food we have introduced into our gut from eating, they will begin to feed on us: protein is protein and the bacteria doesn't distinguish between animal, vegetable or human varieties. When the gut wall is breached, more bacteria are introduced into the body proper where they find a deluge of part-digested food to feed on, courtesy of the autolysis process. The results of this feeding frenzy are unpleasant. This is the second stage of decomposition, described as the 'bloat' stage, where putrefaction takes over and the body is being liquefied from within at an accelerated rate. The body will begin to look and smell awful.

The obvious changes to the body will start with a patch of greenish discolouration on the skin on the lower right of the abdomen that spreads across the body. Meanwhile, the bacteria and enzymes, now moving unhindered through the circulatory system, will be breaking down small capillaries, followed by larger blood vessels, as they go. The skin will develop a mottled appearance.

The by-products of the bacteria's all-you-can-eat buffet are gases including hydrogen, methane, ammonia, sulphur dioxide and hydrogen sulphide. Sulphur-containing compounds are known for their unpleasant smell. Ammonia has a pungency that will be familiar to anyone who has dyed or permed their hair. The smell of rancid fats will start to add to the overall awful mix, as the body's fat reserves are broken down. But these smells have nothing on the appalling stench of cadaverine and putrecine that give decaying bodies their unmistakable and unforgettable reek.

The buildup of gases inside the corpse causes it to swell, particularly around the stomach and gut, but eventually all over as bacteria spread and multiply throughout the body. Flies may have been attracted to the body from a very early stage, just a few hours after death, to lay their eggs in wounds or moist crevices in the body. As autolysis progresses, the increasing stench attracts more insects and other animals that feed and scavenge on decaying material. However, bodies buried in coffins and under several feet of earth are simply not accessible to the flies that are normally the first to arrive and make use of the corpse as a site to lay their eggs. This means it is usually only autolysis and the bacteria already in the body, perhaps with contributions from bacteria in the surrounding soil, that do the bulk of the decomposition work in human burials.

Understandably, bodies are usually buried before they reach this state of decay, or efforts are made to preserve the body a little longer through embalming techniques. The ancient Egyptians were extremely proficient embalmers but the practice was largely lost, certainly in Europe, until the seventeenth century. Even then, it wasn't until the nineteenth century that the practice spread beyond the most important cadavers, such as those of royalty, great statesmen, or high-ranking individuals who had died in battle and had to be transported back to their native soil for burial.

More or less everyone else was prepared for burial by loved ones, or local people who may have specialised in caring for the dead but didn't necessarily have access to effective embalming techniques. The result is that the majority of people were likely to have seen a dead body in the early stages of decay: pale, sunken skin and stiff limbs – the so-called 'fresh' stage. There was a good chance the body would have been interred before rigor mortis had worn off, or the discolouration of the skin had really set in, or the smell had got too bad. The family, friends and those doing the burying would have expected the deceased to stay put and go about its business of disintegration quietly. The corpse would have gone through the remaining stages of decay – bloat, active decay and skeletonisation – underground and out of sight.

In some cultures, after a certain period – months or years – the dead would be re-examined. For some, the soul had not truly departed from the earthly remains until the body had disintegrated, and therefore people obviously wanted to check that their relatives' souls were not still hanging around. In most cases, by the time the body came to be disinterred, it was likely to be in an advanced sate of decay, if not completely reduced to a skeleton. The unpleasant in-between stages would have carried out underground, or in crypts, but out of view. However, things changed when there was a suspected vampire about.

3.3 MISINTERPRETATION

When there was a series of sudden deaths in a community, suspicion often fell on the first person who died in the epidemic. Their body would be exhumed weeks after burial and examined. The examination itself is likely to have been fairly basic. Even on the occasions when there was medical expertise on hand, a hasty autopsy at a graveside, often carried out in the middle of the night, is far from the methodical scientific process that might occur in a well-lit and well-stocked anatomy lecture theatre. Whoever was conducting the examination of the body would not have had much opportunity for close and detailed examination, especially when many of those present would have been keen to dispose of the supposed vampire as quickly as possible.

Any sign that the corpse had not done what it was expected to do could be taken as a sign of vampirism. One of the first things that they were likely to notice when a grave was opened was the smell. The body probably wouldn't have been very fragrant when it was buried but by now, the accelerated decay in the confined space of a grave would have produced a truly appalling stench. The terrible smell emanating from vampires was sometimes commented on and was one way of determining if a well-preserved body was a potential saint or vampire. Holy and unholy bodies are both known for their miraculous preservation but it was important to distinguish between the two. Calmet, who we met in Chapter 1, asserted that a potential saint smelled nice, but vampires did not. In *Dracula*, the Count was noted for his appalling bad breath, which is not surprising after a few centuries lingering in a coffin. The terrible stench Van Helsing and his companions encountered in Dracula's London lair was also taken as evidence that he had recently been at home.

After recovering from the initial assault on the nose, and perhaps waiting for a few moments for the air to clear a little, the remains inside the grave would be examined. The appearance of the corpse would be very different to what it was when it had been buried weeks earlier. First of all, it might not be in the same position as it had been left in. A coffin had to be lifted, carried and lowered into a grave, and all these movements might cause a body to shift position. Even when laid to rest in their grave, dead bodies are surprisingly restless. It may go against common sense and what you might expect to happen, but dead bodies move. As decay processes progress, gas builds up, muscle that was once firm collapses from being digested, bacteria spread and liquids shift within the body.

If signs of movement had aroused suspicions in the minds of those standing round the grave, the obvious next step was to take a closer look. The coffin, and probably the body itself, would have been lifted out of the grave, when it would have become apparent that rigor mortis had worn off and the limbs were free to move again. Several accounts comment on supple limbs and took this as yet another sign of vampirism.

On closer inspection, the body would reveal yet more changes, obvious even in the poor visibility available in a graveyard in the middle of the night. Many accounts of folklore vampires commented on the ruddy complexion of the vampires they had discovered. The last time the face of the deceased had been seen, just before it was buried, it was more likely to have been pale and gaunt. The ruddy cheeks that had developed were thought to be a sign of life, but in fact were the result of the decomposition in the blood vessels and tissues under the skin.

Another sign that was easily misinterpreted was the size and shape of the body. The bloated appearance was not a sign of a well-fed vampire engorged with blood, but was due to well-fed bacteria expelling gases. Swelling would initially be around the stomach and it would look as though the cadaver had just had a good meal. But as the swelling developed over the whole body, it could expand to almost three times its original size and the skin became taught like a drum – a sure sign of vampirism in a corpse in several cultures. As more gas was produced, the pressure would build and may find a means of escape, and an obvious route to the outside world was through the throat and past the vocal chords, causing the corpse to groan.[¶]

Gas is not the only thing that can escape the body. The breakdown of cells releasing their watery contents and mixing with decomposed components of blood and digested food makes a dark brown/red coloured liquid called 'purge' fluid. It can leak into the coffin, and some of this fluid from within the body may have been brought up into the mouth by the gas. Examining a body by a graveside late at night with people jostling to get a view of the body, it is perhaps not surprising this purge fluid was mistaken for fresh blood sucked from the vampire's victims.

As the investigators peered closer, they would have seen more changes in the body. There may have been stubble on the chin, even though the body was buried clean-shaven. Nails may have appeared longer than expected, or fresh nails growing where old nails had fallen off. Contrary to popular belief, hair and nails do not continue to grow after death. Hair and nails are themselves very resistant to decomposition because of the tough materials they are made of. It is the skin around the hair and nails that has receded or shrunk during the decay process because it has dried out. If the investigators were brave enough to look inside the mouth, they would have seen that the same processes had caused the gums to recede, making the teeth appear longer.

[¶]The alternative route was more likely to be blocked by the collapse of the intestines during the decomposition process.

The autolysis process, working away digesting proteins in the body, will also digest the protein of the skin. As more proteins are digested and more cells broken apart, the top layer of skin – the epidermis – will start to slough off, revealing the underlying dermis, which is harder to break down because it contains tough connective tissues that normally give skin its flexibility and resistance to damage. Fingernails and toenails fall off in the process.[||] In the report on Peter Plogojowitz, what was claimed were new nails growing in place of the old ones was simply the underlying nail bed. The underlying dermis layer might have appeared like new skin, but it was not. The fresh and pinkish appearance of the dermis may also have been enhanced because of the bacterial and chemical decomposition going on underneath.

The growing 'evidence' of vampirism may have spurred those at the graveside to take further actions. For some investigators a visual inspection of the body might not have been enough. Anatomists may have wished to inspect the internal organs, and those wanting to dispose of their vampire properly may have been keen to stake the body or remove the heart so they could burn it. Either way, if there were enough external signs of vampirism in a corpse, the body would have likely been cut into or stabbed or staked. The pressure of pushing down on the chest to make an incision could push the accumulated gas through the throat and vocal chords eliciting a final groan or shriek from the vampire. Rummaging around in the chest cavity on a cold night, some would have noticed the body was warmer than expected but again, this was not a sign of life but instead a result of the frenzied activity of the microbes within.

As the number of vampire epidemics increased, it might be expected that there would be a growing familiarity with the process of decomposition as more and more bodies were disinterred in their various states of decay. This could have had several effects on the witnesses present. It could have simply re-enforced their suspicions of vampirism and confirmed that there were plenty of vampires about causing all sorts of problems for the living. The sight of so many bodies in the bloat stages of decay might also have raised doubts that maybe bloat was

[||]Losing nails was such a common occurrence that the ancient Egyptians in their elaborate funeral preparations used special finger caps to hold them in place.

the normal way of things and another explanation was needed for vampirisim.

Even though Europe seemed to be in the grip of a vampire population explosion during the eighteenth century, the actual number of outbreaks was relatively small. Very few people would have been fortunate, or unfortunate, enough to be witness to more than one outbreak. But even a single outbreak could provide several bodies for examination and comparison. In the Peter Plogojowitz case there were ten cadavers that could be compared and contrasted for their signs of vampirism. In the Arnold Paole case there were seventeen! But, if anything, what was revealed during these mass exhumations was how there was no real pattern. The bodies were often in very different stages of decay even though they had all died and been buried within a short space of time in the same graveyard. Why were some bodies better preserved than others?

3.4 WHAT IS NORMAL?

With so many factors at work during the decay process, it can hardly be surprising that there is enormous variation. Many different scenarios can in fact be normal, or at least explicable by natural processes before resorting to the supernatural. 'Unusual' doesn't mean unnatural.

Take the single characteristic of rigor mortis. How quickly it appears in a body can depend on what the person was doing immediately before death. There will be huge variation between individual cases, from people falling down dead fixed in the position they were in when they took their last breath, as has been observed on battlefields or with drowning victims, to bodies that never develop rigor mortis at all. A lot of physical exertion in the dying moments – because someone is having a seizure, fighting off an attacker or thrashing around to save himself or herself from drowning – can cause rigor to set in very quickly because much of the ATP generated by respiration has been used up.** Poisoning with strychnine can hasten the onset and disappearance of rigor mortis, whereas carbon monoxide can delay it.

**This sudden onset of rigor mortis is sometimes called cadaveric spasm.

Another factor is changes in temperature, which can accelerate or delay both its appearance and disappearance.

Temperature will also affect the work of bacteria and enzymes. Bacteria and enzymes work best at around 37 °C (98.6°F). Their activity slows as the temperature drops, and can be stopped completely at high temperatures as bacteria are killed and enzymes denatured. This will dramatically affect the length of time taken for the body to go through the autolysis process. Under 10 °C (50°F), or above 40 °C (104°F), putrefaction is greatly inhibited. Bodies buried in winter might still be looking fairly fresh by spring. Whereas the increase in temperature over the spring and summer months will help to accelerate the decay process. The Plogojowitz and Paole cases both occurred over the winter months when there was the best chance of bodies buried in the cold ground being preserved.

However, the cold weather doesn't explain how bodies all buried around the same time and in the same graveyard could have been in such differing states of decay. One explanation might be how the bodies were buried. Could some of Plogojowitz and Paole's richer victims have been buried in expensive coffins? A solidly built coffin can keep out the bacteria from the surrounding soil much more effectively than a simple winding sheet, which may have been all the poorer victims were able to afford. Some remains may have been placed in family tombs, while others were buried in less salubrious sites. Some may have been buried deeper in the ground where it is colder and more likely to be damp, than shallow burials that will experience more variable temperatures. If the grave is very shallow, the smells emanating from the decaying body can attract insects and larger animals to scavenge on them.

Insects and other animals will generally speed up the decomposition process. However, in extreme circumstances nocturnal scavenging animals, such as wolves, may try to dig up bodies they can smell not too far under the surface. But even large animals may not be able to drag a whole adult human body out the ground. In the past, poor church wardens have arrived at a graveyard early in the morning to find a body half hanging out of its supposed resting place like something out of a zombie apocalypse movie.

Water is another important factor. Bacteria need water to carry out metabolic process, just as humans do. The ancient Egyptians deliberately dried out their dead to preserve them. Bodies buried in dry soil can dessicate. Bodies buried in lower ground might be more susceptible to ground water seeping in. And, though water would normally be expected to accelerate decomposition, especially if that water washes more bugs and bacteria into the body, in some circumstances water can help preserve a body.

One of the many ways water can interact with a dead body, and the bacteria colonising it, is to hydrolyse (literally meaning 'split apart with water') fat molecules. Adipose tissue in the body contains fats mostly in the form of triglycerides. During decomposition, these triglycerides can have their three glyceride units broken off by water to form free fatty acids. Hydrogenation processes (adding hydrogen atoms) can then transform these fatty acids into saturated fats, which can then form adipocere – a waxy solid that looks a bit like hard soap. Adipocere, also known as grave wax, normally starts to develop around a month after burial, should be visible to the naked eye after about three months, and, given enough time, can cover the whole body.

Once formed, adipocere is remarkably stable and can preserve the shape of the body and organs for years. In a well-known case from the UK in 1911, two boys, the Higgins brothers, were murdered by their father and their bodies thrown into a flooded quarry. The bodies were found floating, *two years later*. In the intervening time, adipocere had formed throughout almost their entire bodies. Their remains were so well preserved, forensic examination was able to not only identify the two boys, but also provided enough evidence to convict their killer.

The Higgins brothers case is extreme, but you can see how grave wax could give an eerily undecomposed appearance to a body. The process can occur in a wide range of burial situations, but there are some that retard and others that increase its formation. Water is needed for the essential hydrolysis step and there is usually enough water in the body itself for adipocere to occur, but moist or wet surrounding are likely to increase the rate of formation and extent of adipocere. However, one of the most important factors is oxygen. The more air circulating around a body, the more rapidly it will decay thanks to the work of aerobic bacteria. Bodies sealed up in coffins and/or buried deep underground are more likely to support the anaerobic bacteria than can transform body fat into adipocere. Bodies in coffins may be sealed off from the bacteria in the surrounding soil but there are plenty already in the body to carry out the work of forming adipocere. Many bacteria are sensitive to pH, and those involved in adipocere formation work best in slightly alkali conditions.[††] However, a very alkali environment, such as burials in lime, can preserve the body by chemical dessication.

Quite apart from the burial environment, there is the body itself. Individuals vary, as do the manner of their deaths. Did an individual die of a bacterial infection? Were there open wounds that would allow bacteria and flies easy access to the interior of the body? Were they a small child or a full-grown adult? Was the person old or young? Fat or thin? All of these can contribute to the way the body decays.

On top of the huge variations between individual bodies and burial sites, there are also regional variations that can contribute to misunderstandings and misinterpretations of what are actually perfectly ordinary patterns of decay. European folklore vampires are expected to be well preserved *and* bloated or red, which is actually a contradiction, because the bloating and redness are signs of decay. However, Chinese vampires are very different – they are easily identified by their covering of pale green hair, perhaps due to mould or fungus growing on the skin of the corpse.

And, what is normal for one region might be considered highly suspicious elsewhere. Take the Greek island of Santorini as an example. The island was well known for its vampire problem, something that may have more to do with its geography than anything else. The island is volcanic, and through the combination of heat from the volcano below, the hot climate and arid conditions, it has been suggested that the soil the local people would have been buried in may contain fewer bacteria for decomposition. But, as we have seen, the body itself is full of bacteria that can do the job without help from outside, it just might take a little longer.

Sometimes, local customs may have contributed to stories of vampires in the region, such as in Styria. In the nineteenth century, in this mountainous region of Austria, some of the locals had the habit of eating arsenic. Working at high altitude or doing hard

[††]Very acidic environments can also preserve, as seen in the many examples of bog bodies discovered in remarkable states of preservation, despite being in the ground for centuries or millennia. This is because most bacteria do not tolerate very acidic conditions well.

physical labour in deep mines, they said that eating a small amount of arsenic, perhaps a lump the size of a grain of rice two or three times a week, helped them breathe. The arsenic also bulked them up a little bit, making the men more muscular and attractive. Women also took arsenic because it gave them beautiful curves. Everyone who crunched arsenic between their teeth had fantastic, flawless complexions and glossy, lustrous hair.

The apparent good health of the arsenic eaters of Styria was an illusion. Sub-lethal quantities of arsenic can stimulate the production of red blood cells in the body, but this is not a good reason to take it. The bulking up and beautiful curves were the result of oedema, the accumulation of fluid in the tissues, and not a good thing. Their complexions were improved because the arsenic was killing the bugs and bacteria that would cause spots and blemishes. Their shiny hair was from arsenic bonding to the sulphur atoms in keratin molecules that are part of the structure of strands of hair.[‡‡] All of the attributes they found desirable were in fact early signs of poisoning. Though each dose of arsenic was not enough to kill on its own, arsenic is excreted only very slowly, and so can easily build up inside the body to lethal amounts. What has this got to do with vampires you might ask?

Well, everything became clear after these arsenic eaters eventually died, whether from their addiction to a toxic beauty treatment or other causes. Because there was limited land available for burials, the tradition in Styria was for internments to be only a temporary situation. Twelve years after the funeral, the body would be disinterred and the bones moved to a crypt or charnel house, leaving a vacant plot. But, when it came to removing the bodies of arsenic eaters, they were very different. Their bodies had not been reduced to skeletal remains. The arsenic in their body had killed off the bacteria that would cause much of the destructive decomposition process. Bodies of arsenic eaters were so well preserved that they were recognisable by friends and family. Before it was realised that unusual dietary habits were the reason for their remarkable preservation, these bodies may well have been the source of vampire stories in the region.

[‡‡]It leaves a fantastic chemical history of an individual's exposure to arsenic as the hair grows out from the head. Analysis of hair is how forensic toxicologists can determine a reliable history of poisoning in criminal cases.

3.5 'THE TRUTH WILL OUT'

In the cold light of day and modern scientific understanding, the notion of dead bodies being mistaken for the undead can seem a little silly. The description of the remains of Peter Plogojowitz and his victims contains clear evidence that they were going through the perfectly normal stages of decay, though perhaps a little slower than usual. But in 1725, when Peter was dug up, knowledge of bacteria and their role in decomposition, as well as the chemical reactions necessary for autolysis and the formation of adipocere, were not known. The vast differences in the progress or appearance of decay between different bodies must have appeared extraordinary to those huddled around a grave in the middle of the night. These people, already fearful of the epidemic that was killing their families and neighbours, can hardly be blamed for getting things wrong.

However, these people hadn't started digging up their deceased relatives on the spur of the moment. Disease was common and deaths would have been expected. There must have been other signs that indicated to them that the cause of their troubles was not part of the ordinary run-of-the-mill life and death in the eighteenth century. What other signs prompted these people to go to their graveyards with a shovel and a stake?

Peter Plogojowitz had been singled out as a vampire because he had been seen wandering around the village after his death. This may seem difficult to explain scientifically, but there is a rational explanation. Peter's recently bereaved widow cannot be blamed for being preoccupied about her husband. It would not be at all surprising if she dreamt about him after his death. As stories of Peter's apparent return spread through the village, it would not be so strange if he was dreamt of by others who had known him in life. All of the vampire's predatory attacks, that were so carefully described and recorded, have the victim attacked in their bed while they were asleep. Their dreams were perhaps only made more vivid and fevered because of the illness they had contracted. However, the loss of blood so often seen in the victims of alleged vampire attacks cannot be explained by nightmares. The vampire's insatiable blood lust will be the subject of the next chapter.

CHAPTER 4

Blood

'The Blood is life ... and it shall be mine.'

Bram Stoker, *Dracula*

The Countess Erzsebet Báthory de Esced is also known as 'The Blood Countess'. She lived in Hungary at the turn of the seventeenth century and had a reputation for being very cruel to her staff. When officials went to her castle to investigate the rumours, they found the truth was much, much worse. Inside the castle were dozens of members of the Countess's household, all young women and girls, dead or dying. It was said that she bathed in their blood to keep herself looking young and beautiful.

Báthory, a distant relative of Vlad Dracula, was born in 1560. She was engaged at ten and married at fifteen to Ferenc Nádasdy, a wealthy and influential nobleman. As part of his wedding present to his new bride, he gave her his household, the Castle of Csejte, along with a country house and seventeen neighbouring villages. Even so, Báthory was of higher social standing and refused to change her name.

Around 1601, Nádasdy fell ill with pains in his legs and by 1603 he was permanently disabled. He died in 1604, leaving his wife and heirs in the care of György Thurzó, Palatine of Hungary, and the man who would lead the investigation into Báthory's crimes.

The rumours had started around 1602. In one version of events it all began when one of the maids was careless when

Vampirology: The Science of Horror's Most Famous Fiend
By Kathryn Harkup

Published by the Royal Society of Chemistry, www.rsc.org

combing the Countess's hair. She hit the girl giving her a bloody nose and some of the blood splashed on the Countess's arm. When she washed the blood off, the skin underneath appeared to be younger looking. Another account says that her blood lust stemmed from childhood. When she was very young the Countess suffered from seizures, perhaps epilepsy. A common sixteenth century treatment for the condition was to drink the fresh blood of an executed criminal.

Whatever the cause, she became increasingly violent in later life. Initially the violence was directed towards her staff, but she later moved on to the daughters of the nobility that had been sent to her. It was quite normal at the time for children of the aristocracy to be sent away to great households to learn skills appropriate to their social status. But when these girls were supposed to be learning needlepoint and social niceties, they were instead being tortured and killed.

Word of their appalling treatment spread and an investigation was launched. Around 300 testimonies were collected before Thurzó went to the Castle of Csejte. The evidence he found there led to Báthory's arrest, along with four members of her staff who were said to be her accomplices. Two trials were conducted. At the second trial, a register was submitted as evidence listing 650 names, or supposed victims, in the Countess's own handwriting. She and her accomplices were found guilty.

Three of the Countess's staff were executed, but Báthory was spared the ultimate punishment because of her noble birth. Instead, she was bricked up inside a suite of rooms inside her own castle. She remained there for three and half years, surviving on food pushed through a gap in the wall. One night she complained to her gaolers that her hands were cold, but she was told it was nothing and to go to bed. Her dead body was found the next day (see Figure 4.1).

Many of the stories surrounding the Countess Báthory, of her biting her female staff and bathing in their blood, emerged after her death and were undoubtedly either exaggerated or entirely invented. Báthory has gone down in history as the most prolific female serial killer, although there is little hard evidence of what really went on in her castle. Crimes were certainly committed; the official tally of victims was 80, still a horrific number but nowhere near the alleged 650. In the testimonies that were collected, and

Figure 4.1 Portrait of Countess Erzsebet Báthory. This painting is a sixteenth-century copy of the 1585 original that is now lost.

during the subsequent trials, there were allegations of many things, including witchcraft, but there was no mention of the Countess bathing in her victim's blood. Centuries of gossip and sensational stories of her reputed blood lust has left her with a reputation as a real-life vampire that has proved impossible to shake off. Báthory's interest in blood may have been extreme but this life-giving substance has fascinated humans for millennia.

4.1 BLOOD, BLOOD, GLORIOUS BLOOD

Blood has a special significance from a cultural, historic and scientific point of view. This red liquid that flows through all of us has been used for religious sacrifice, as food, medicine and has even been thought of as a potential fountain of youth. Though not all ancient or folklore vampires drank blood, it is a key feature of vampires in literature and film. A preoccupation with blood is one of the first characteristics someone will list when asked about vampires.

That blood is essential to human life has always been known, but over the centuries it has been imbued with almost mystical reverence and special properties. As scientists have learnt more about this wondrous liquid, it has lost none of its status as a truly remarkable substance. It supplies vital oxygen to our cells, distributes food, warmth and hormones, removes waste and acts as a primary defence against invading pathogens. And there is still much more to discover. Blood has more than lived up to historic expectations.

Could this incredible liquid really have the powers it appears to possess for a vampire? Could it sustain someone in an undead existence? Does is really have the power to nourish, cure disease and keep you looking young and beautiful? The answer is both yes and no.

Blood is not a single substance, it is a mixture of different components carrying out many vital roles within the body. The major constituents of blood are: red blood cells that pick up oxygen from the lungs; white blood cells that are a key part of our immune system; platelets that form clots to stop bleeding and form a barrier to the outside world; and plasma, the fluid that all the components float around in as they move through the body. Each component is of potential benefit to the vampire. Blood, and some of its many components will be mentioned frequently in the coming chapters so it is worth spending a little time exploring this marvellous substance in some detail.

The biggest component of blood is plasma, accounting for 54.3% of the volume of blood, closely followed by red blood cells, which make up around 45%. White blood cells take up most of the remaining 0.7% of the space. Platelets, even though they greatly outnumber the white blood cells, are tiny and only take up a very small amount of the overall volume.

The most familiar feature of blood are the red blood cells, the cells that do the all-important job of transporting oxygen from the lungs to the rest of the body. Oxygen is one of the raw materials we need for the respiration process that produces energy. The other ingredient for respiration is sugar, which we can stockpile in our body as fat. We cannot store oxygen, meaning we need a constant supply. It is red blood cells moving oxygen from our lungs to our cells that keep us alive from one minute to the next.† There are trillions of them speeding through our circulatory system‡ and evolution has refined them to optimise the amount of oxygen they can carry. Their shape, like a squashed doughnut, gives them a bigger surface area than if they were simply spherical, allowing oxygen to diffuse as easily as possible into the cell. The cells are also flexible so that they can squeeze through the smallest capillary and there are yet more modifications on the inside.

Each red blood cell is packed with haemoglobin, the molecule that actually binds the oxygen from the lungs and holds on to it during its journey through the body. Haemoglobin molecules take up about 35% of the cell's total volume. This may seem like a small proportion of the cell's contents, but it is a lot when compared with other cells in the body. Most cells contain around 70% water, leaving the remaining 30% of the available space for all the structures and molecular machinery that makes the cell function properly. To make 35% of the available cell space just for haemoglobin, red blood cells have sacrificed several components – they do not have a nucleus or mitochondria or several other structures that would normally be essential for cells to function properly – and yet they work brilliantly at the job they are needed for.

Haemoglobin is an extraordinary protein made up of four units that each bind one oxygen molecule. It protects the body from the potentially harmful effects of letting a reactive molecule like oxygen just roam around unguarded and uncontrolled. The units work cooperatively to maximise oxygen uptake. Once the first oxygen molecule is bound, there is a slight shift in the physical structure of the protein that makes it easier for the second to bind and so on. The oxygen molecules remain bonded to the haemoglobin until the red blood cell reaches a part of the body that is low in oxygen. The low concentration of oxygen outside the cell encourages the haemoglobin to release its oxygen where it is needed. Haemoglobin also helps transport around a quarter of the waste carbon dioxide produced in respiration back to the lungs so it can be breathed out. It is a fantastically efficient system.

---

†Quite literally, as it takes approximately 60 seconds for a red blood cell to complete a circuit around the body.

‡At roughly two to three miles an hour.

The relentless absorbing and releasing of oxygen carried out by red blood cells, whilst travelling thousands of miles in endless circuits around the body, soon wears them out. Red blood cells have a lifespan of around 100 to 120 days. They cannot replicate the way other cells can because they have no nucleus and therefore no DNA instructions for making more of themselves. Instead, old cells are broken down and the elements are recycled while new cells are being produced in the bone marrow. Each new cell takes seven days to produce, hardly surprising given the complexity of the task, but it is being carried out on a massive scale. In healthy adults, every single second of every single day, two to three million new red blood cells are produced.

The regenerative properties of blood are remarkable and allow people to donate blood regularly without draining themselves completely. This might have been known for centuries, as in the past, when blood-letting was used as a medical therapy, there was always more blood to be found when a patient returned for further treatment. It also means a vampire can, theoretically, keep their food supply going indefinitely, as long as they don't get too greedy.

While red blood cells are important for survival from one moment to the next, other components in the blood play important roles in keeping us healthy over longer periods of time. Protecting us on a slightly longer timescale are the platelets. These tiny plate-shaped cells are what form clots. Damaged blood vessels send out chemical signals that attract the platelets. These cells then grow tendrils so that they can stick to the damaged part of the vessel. As more platelets stick, fibrin (long strands of protein normally dissolved in the plasma) also become tangled up and trap more cells, including red blood cells, which form a solid barrier that stops blood leaking out of the blood vessel. Clots, or scabs, formed on external cuts also prevent germs entering the wound and causing an infection.

The vampires of films and books are well known for their ability to rapidly heal themselves. Only extreme wounds, like a stake through the heart or decapitation, are enough to stop them. Their clotting process must be formidable. However, forming large clots and scabs quickly and easily also has a downside. People with very high numbers of platelets in their blood are more susceptible to forming clots where they are not wanted (thrombosis), for example, in the veins of the legs.

These clots can become detached and travel through the blood vessels and lodge somewhere else, blocking the flow of blood and potentially leading to heart attacks and strokes. But maybe vampires, who don't have a heartbeat or bloodflow, don't need to worry about thrombosis.

Blood clotting is the first line of defence against pathogens entering the blood system, but it is not foolproof. The circulatory system is brilliant at delivering oxygen, heat and hormones to every corner of the body, but that also means it has the potential to deliver harmful substances or organisms just as easily. Humans, and vampires, need further protection from harmful invaders and that comes in the form of white blood cells. They may take up less than 1% of the blood's volume, but the play a huge role in the body's immune defences.

As far as the immune system is concerned, everything in the body can be divided, very broadly, into two categories – self, and non-self. Cells within the body have tiny identifiers in the form of proteins, called antigens, which are present on the cell surface. White blood cells roam around the body, chemically examining everything they encounter, and the antigens it expresses. If the object under scrutiny displays antigens it knows to be self, it is ignored. If it is not recognised, or is recognised from a previous infection as non-self, an immune response is launched and the object, be it a bacteria, or virus, is attacked and destroyed.

There are many varieties of white blood cell involved in identifying and destroying invaders but they can be grouped, very approximately, into lymphocytes and phagocytes. The lymphocytes produce antibodies, y-shaped proteins that stick to the surface of foreign bodies such as bacteria and viruses. These antibodies perform several functions. First of all, inactivation of the intruder, because a virus covered in antibodies cannot enter a cell to infect it. Secondly, antibodies act as chemical flags to recruit other parts of the immune system to home in and attack whatever it is the antibodies are stuck to. The antibodies instruct the phagocytes to engulf the whole virus or bacterium, complete with its hedgehog-like decoration of antibodies, and digest it.

Humans maintain a huge stock of different antibodies in their immune system as a kind of reference library, so that a wide range of pathogens can be targeted and destroyed. This library is constantly being updated as the body is introduced to more pathogens over its lifetime and new tailor-made antibodies are created and remembered. It is thanks to this system of remembering past infections that vaccines can prime a person's immune system. By introducing fragments of a virus, weakened or inactive forms of a microbe, the body can produce antibodies in preparation for an infection of the bacteria or virus proper. It is also why people can catch some diseases, such as chicken pox or measles, only once in their lifetime.[§]

A vampire's immune system must be phenomenally good to keep them healthy throughout their extended lives, especially when their lifestyles would seem to be particularly high-risk. Sleeping in a grave is far from sanitary, though presumably coffins can be kept clean, at least up to a point. Crawling through grave dirt to get up every night will expose a vampire to a huge number of pathogens, no matter how much they scrub and sterilise their coffin. Their choice of food is also high-risk. Just to get their food, they have to get very close to their victim, potentially exposing them to any number of contagious diseases and the blood itself can harbour yet more pathogens and parasites.

On the other hand, blood's life-giving properties have been known for millennia, even if the mechanisms were anything but clear. It didn't require a detailed understanding of oxygen binding and respiration to know that losing large amounts of blood resulted in death. It was only a short step from knowing blood equated with life to believing blood could also heal. Even before germ theory, or an understanding of what white blood cells actually did, blood was seen as a way to fight off disease and was used as a medicine.

4.2 BLOOD AS MEDICINE

Blood has had a central role in medicine for centuries. Up until the seventeenth century, human health was understood in terms of the four 'humours'. A correct balance of black bile, yellow bile, phlegm and blood was thought to be the reason for good health. If any of these substances were deficient or present in excess it was thought to cause illness. Consequently, people were fed

[§]Though often thought of as mild childhood diseases, both chicken pox and measles can have serious, life-threatening complications. Vaccinations prime the immune system, making infections milder and hugely reducing the chances of dying.

certain foods that were believed to increase the amount of one humour or another, or they were made to vomit, or given laxatives, to get rid of excesses. And of course, people were bled, not to feed vampires, but to remove heat from a body if someone was feverish and to relieve a whole host of other symptoms of illness. The process was called phlebotomy, and it continued in medical practice until well in to the nineteenth century.[¶] In fact, healthcare professionals who collect blood samples for testing in hospitals and clinics are still called 'phlebotomists'.

Blood was not just removed from the body, it was also administered to patients in the hope that it would confer some benefit. The story of Erzsebet Báthory, the Blood Countess, being treated for her childhood seizures by drinking blood was not unusual,[‖] it was in the past a popular treatment for epilepsy. Blood-drinking epileptics have been written about in Germany, Sweden and Austria, and blood was thought to have healing properties beyond epilepsy.** The blood in question was obtained from criminals during their execution. When an execution was announced, people would rush to the scaffold to place cups and bowls on the wooden platform to catch the blood as it gushed from the wounds, though some made do by dipping their handkerchief in the spilt blood.

Ingesting blood, from executed criminals or anyone else for that matter, had no therapeutic benefit whatsoever, but at least it did less direct harm (to the recipient anyway). Blood certainly has medical benefits and has saved countless lives, but through transfusions. Though people had been ingesting blood for centuries, no one tried injecting it into veins until the seventeenth century, and then not as a cure for epilepsy, or to counteract blood loss, but as a treatment for mental health problems.

4.3 BLOOD TRANSFUSIONS

The four humours theory was all-encompassing. Along with physical health, the humours were also thought to influence your

---

¶ To little or no benefit, it should be added.

‖ What *is* unusual is her later career as a serial killer; it was not a normal consequence for the many people who suffered from epilepsy and the many more who used blood as medicine.

**Drinking blood is not a cure for epilepsy or any other disease; please don't self-medicate.

mental health, mood and personality. Phrases we still use today when talking about people who are in a good or bad humour are hangovers from earlier times, when they had a more literal significance. It therefore seemed logical, at the time, that blood could also be used to restore balance to someone's mental health.

Despite the enthusiasm of people in the seventeenth century for draining blood, they were considerably more cautious when it came to injecting it into people. The first experiments with transfusion were conducted in 1666 on dogs. Robert Boyle, a prominent natural philosopher of his day, along with Richard Lower, one of Oxford's finest physicians, opened the jugular vein of one dog and inserted a pipe, which was then connected to the jugular vein of a second dog. Unfortunately, the blood clotted in the pipe and both dogs died.

Lower took up the experiments again, but without Boyle and with a modified method. One problem with the earlier experiment was that the pressure in the two veins of the dogs would have been the same, and blood would not have been able to flow from one to the other. In his new experiment, Lower instead connected a silver pipe between an artery in the first dog and a vein in the second dog. The higher pressure in the artery forced the blood through the pipe and into the second dog without clotting. Lower almost bled the first dog to death before transfusing it with a fresh supply. At the end of the experiments the dog apparently leapt off the table, licked Lower enthusiastically, and rolled on the grass to clean its fur.

The Royal Society, hearing of Lower's success, took up the baton and drained calves' blood into sheep, and lambs' blood into foxes. Remarkably, the sheep survived. The foxes died. Undeterred, when Lower took over the transfusion experiments at The Royal Society, he decided to take the bold step of transfusing blood into a man. In keeping with the humoural theory, Lower hoped that an injection of calming lambs' blood would soothe the troubled minds of the insane and 'improve his mental condition'. He tried to persuade Bethlem Royal Hospital, London's foremost hospital for the insane at the time, and the origin of the word 'bedlam', to surrender one of its patients. The hospital refused to let any of their patients take part in such a ridiculous experiment. Instead, Lower found a tramp named Arthur Coga in a church congregation, a man described by Samuel Pepys as 'cracked a little in his head'. Coga was in fact a Cambridge-educated Bachelor of Divinity but he had turned to drink. Lower promised him 20 shillings (£1) in return for allowing twelve ounces of sheep's blood to be let into his veins. Coga agreed.

Silver pipes were used to connect an artery in a lamb's neck with a vein in Coga's arm. Miraculously, he survived. The experiment would have been hailed as a tremendous success if Coga hadn't taken his twenty shillings and spent it in a pub getting drunk and raving about Lower's experiment. It was hardly a ringing endorsement for the calming effects of lambs blood, but the procedure itself had worked, and with no apparent harm to the human subject. It prompted others to speculate about the therapeutic possibilities of blood transfusions, from calming madmen to rejuvenating the old. The excitement soon came to an end when the experiments resulted in the deaths of their human volunteers.

Similar experiments had been conducted in France around the same time as Coga's transfusion, by Jean-Baptiste Denys. In 1667, Denys also transferred blood from animals to human subjects and was initially successful. His first two volunteers survived but probably only because so little blood was actually transferred. This was not the case for Denys's last subject, Antoine Mauroy. Mauroy, like Coga, was considered a madman and therefore Denys selected calf's blood for the transfusion. At first, all went well but a second, larger transfusion was very different. As the blood entered his arm Mauroy's pulse rose and he started to sweat. Unknown to Denys, Mauroy was going into haemolytic shock, effectively a massive allergic reaction to the foreign blood. He vomited and later 'made a great glass of urine, of a colour as black as if it had been mixed with the blood of chimneys.' The soot colour was caused by chewed up red blood cells that had been destroyed by the immune system. Though Mauroy survived long enough for Denys to claim the experiment was a success, he soon died and Denys was disgraced. Blood transfusions were promptly banned across Europe and research effectively came to a complete halt for the next one hundred and fifty years.

Interest revived in 1818 when James Blundell, an English obstetrician, carried out a successful blood transfusion by injecting blood into a woman haemorrhaging after giving birth. The donor was her husband, and both survived. Blundell carried out more transfusions during his career but only ever used human blood.

He worked quickly to avoid clotting but, of the ten transfusions he carried out in his career, only five of the recipients survived.[††] It was progress, but transfusions remained highly controversial in medical practice because of the very real dangers to the patient.

Blundell was using blood transfusions as a last resort, when severe blood loss was almost certainly going to kill the patient. Why donated blood could save some but kill others was a mystery. The idea of blood compatibility was starting to be appreciated, but the biological reasons behind it were a mystery. This was roughly the state of knowledge when Bram Stoker came to write *Dracula*. Blood transfusions were controversial and full of scientific and social taboos.

In the novel, Dracula has been drinking Lucy Westenra's blood on a nightly basis. Small losses of blood can be replenished by the body; the body's blood volume will be back to normal twelve to sixteen weeks after a standard donation.[‡‡] Regular feeding by a vampire, however, is not the same as safe, measured, occasional blood donation. The vampire doesn't need to drain the body completely in order to threaten the life of their victim. If the vampire's greed outpaces the body's ability to replenish the lost blood there are serious consequences for the victim's health.

Humans contain, on average, nine pints of blood (around five litres). As well as oxygen, the heart needs a certain blood pressure to keep pumping. Losing more than 20% of your blood volume means the pressure drops to a point where the heart will struggle to move it around your body. If blood is not being pumped to the brain, the brain becomes starved of oxygen and death will occur in a few minutes. Losing 5% cent of your blood volume per minute, through injury or a vampire guzzling on a litre of your blood, would mean death in four minutes.

In *Dracula*, Lucy Westenra's blood loss must be smaller, as she survives a lot longer than four minutes. The first attack on her occurs on 11 August, and her health declines steadily over the following four weeks. The effects of regular blood loss were exactly what you would expect, and she displays many of the characteristic symptoms of anaemia – 'the roses in her cheeks

[††]Admittedly, two of those who did not survive were dead already – the transfused blood failed to resurrect them.

[‡‡]A standard blood donation in the UK is 470 ml, about a pint, which is roughly 13% of the body's total volume, or 'nearly an armful'.

are fading, and she gets weaker and more languid day by day; at night I hear her gasping as if for air'. Lucy's suitor, Dr Seward, was understandably concerned, and thought he could bring his medical expertise to try and solve the problem. An examination of her blood showed no signs of anaemia and so he was stumped as to the cause of her bloodlessness.

Anaemia is a decrease in the number of red blood cells or haemoglobin in the body. It can also be defined as a decreased ability to transport oxygen. There can be a number of reasons for this, only one of which is blood loss. Dracula evidently had tidy table manners and didn't spill the blood he was drinking, leaving no tell-tale blood stains. When there is no obvious bleeding, the first step in diagnosis is usually to do a red blood cell count. Dr Seward would have been able to do this with the blood sample he took from Lucy.

When blood is lost from the body, for example, through a regular donation, roughly two thirds of what is taken is water, which can be replaced by the body within the day, and it's why donors are encouraged to drink lots of fluids afterwards. Red blood cells and other components take much longer to replace and so Lucy's blood should have shown a low red blood cell count. The components of the blood itself should have looked perfectly normal, there would just be fewer of them floating around in the plasma. Perhaps Stoker was confused over the details and expected anaemia to show some alteration to the cells. Or, perhaps he didn't realise that much of the volume of blood that his main character had drunk would be replaced by water from the rest of the body. But this is splitting hairs. The point was that Dr Seward could see no obvious cause for Lucy's condition and looked for a second opinion.

By the time Seward brought his mentor Professor Van Helsing in to consult on Lucy's case on 7 September, her condition had deteriorated to the point that her life was in immediate danger – 'She will die of sheer want of blood to keep her heart's action as it should be.' A blood transfusion was carried out, with Arthur Holmwood, another of Lucy's suitors, as the donor. Such drastic and potentially lethal action would only have been contemplated in the most dire of emergencies.

Why some patients died and others survived transfusions wasn't understood until 1901, four years after the publication of *Dracula*, when Dr Karl Landsteiner discovered that blood came in four main types. It was Landsteiner who determined the basic categorisation of blood into A, B, O and AB.

Two types of antigen can be expressed on the outside of red blood cells, resulting in blood groups A (one type of antigen), B (a second type of antigen), AB (both antigens present), and O (no antigens present). If someone with type A blood were to be given type B blood in a transfusion, the recipient's white blood cells would examine the B antigens on the surface of the introduced red blood cells and, not recognising them as self, would launch an immune response. Hence, people with type O blood are known as universal donors. Anyone can receive this type of blood because it has no A or B antigens to identify it as self or non-self and won't be attacked by white blood cells.

Since 1901, more antigens have been identified. By the 1950s, 25 red blood cell antigens were known, and there are now over 300 that have been identified. The seventeenth-century experiments transfusing blood between dogs had been misleading because dogs' blood does not express the same antigens as human blood and can be transferred safely between different breeds.

In *Dracula*, Lucy and Arthur both survived their first blood transfusion, but Lucy's condition continued to deteriorate. Further transfusions were made using four different donors, all of them unrelated to her. She was extraordinarily lucky to survive the procedures, but even repeated donations could not keep up with Dracula's seemingly insatiable thirst. What could the blood be doing for Dracula that he would drink so much of it?

4.4 BLOOD AS FOOD

Blood drinking is far from unique to vampires. Many animals feed on the blood of other animals, from fleas to mosquitoes and, of course, vampire bats.[§§] Humans can't be left out of the blood-drinking either. Plenty of humans eat blood in the form of black pudding, haggis or, in some cultures, straight from the cow and mixed with milk as a nutritious drink.[¶¶] What distinguishes

[§§]There are an estimated 40 000 different species of parasite that drink human or animal blood.

[¶¶]Blood can be an excellent source of protein for people who have restricted access to it. Occasionally taking small amounts of blood also does no harm to the cow.

vampires from human blood-feeders is that humans are (rarely) drinking blood fresh from other humans.‖

The majority of vampires seem to shun conventional human food and feed exclusively on blood. From an energy point of view, it is not a very sensible choice if vampires have anything like the metabolism of humans. Simply to supply him with sufficient calories, Dracula would need to drink four litres of blood every night – roughly 80% of an adult's total volume. Given some of the superhuman abilities Dracula appears to possess, he would likely need a much higher calorific intake, but let's stick with the human analogy for the time being.

Calories are one thing, what about the nutritional value of blood? Within those four litres of human blood, there would be less than four grams of carbohydrates and twenty grams of fat – far less than the recommended daily amount of both nutrients. By contrast, protein levels would be in huge excess. Four litres of blood contains around 800 g of protein, more than thirteen times the recommended daily intake. More worryingly, there would be approximately 36 g of salt in Dracula's nightly feed, six times the recommended daily amount. It would probably leave him with a raging thirst that would lead to dehydration if he didn't quench it with several litres of water to wash down his dinner. High-salt diets are associated with high blood pressure and therefore increased risks of heart disease and strokes. In extreme cases, it can lead to hypernatraemia, in which too much sodium, from the sodium chloride, can cause weakness, headaches and even convulsions.

If salt poisoning doesn't put you off an all blood-diet then there are more nutritional disasters awaiting you. When it comes to micronutrients, blood fails to provide the right balance of vitamins and minerals that are essential to a healthy human life. Take vitamin C, for example; can blood count as one of your five-a-day? In short, no. The low levels of vitamin C would likely lead to scurvy.*** A lack of calcium in blood would also be a concern, especially if you need to keep your fangs nice and strong. Another serious concern would be iron, because there would be too much, rather than too little.

‖There will be more on this in later chapters.

***Symptoms include tiredness, weakness, joint pain, swollen and bleeding gums, as well as skin that bruises easily.

The human body contains around 4 g of iron and most of that is found in the haemoglobin in red blood cells. To get this iron, we eat it in food, but the iron has to be in a form that the body can utilise – and that is a rare thing. To deal with this scarcity, the body is very good at storing reserves of iron and excreting very little (around 1–1.5 mg per day) and our diet only needs to contain enough iron to replace this loss. Dracula's repast of four litres of blood would contain over a gram and a half of iron. Humans normally absorb only around 10% of the iron in their food and absorption is helped by vitamin C. The lack of vitamin C in a vampire's diet would reduce the amount of iron absorbed into the body, but not by anywhere near enough. An all-blood diet would very likely lead to iron poisoning, or haemachromatosis.[†††]

Iron poisoning would be a serious side effect for any human who tried to live on an exclusively blood diet, but both scurvy and haemachromatosis take time to develop. Drinking large amounts of blood usually makes people sick quite quickly and would put off most from pursuing it as a long-term dietary plan. Perhaps vampires, being dead, are less concerned about balanced diets, but they still need food to support their continued undead existence. Humans may have problems surviving such a restricted menu but there are other animals, even mammals, that seem to thrive on blood.

Vampire bats, unsurprisingly, get their name because of their feeding habits.[‡‡‡] They survive and thrive on an all-blood diet, mainly taken from cattle, but they have had to make many biological compromises to achieve this. Just getting their meals takes stealth and a gentle approach so as not to wake the sleeping cows. They have the longest thumbs of any mammal (for their size), meaning they can walk and stalk their prey by using their thumbs as feet, or run away fast if they are disturbed.

In the past, it was believed that vampire bats used the gentle beating of their wings to send their victims into a trance-like sleep. In reality, it comes down to a lot of training and support. To survive as a vampire bat requires a high level of sophistication

[†††]Symptoms include fatigue, weight loss and joint pain. The primary treatment for iron poisoning is phlebotomy, or blood letting, one of the very few medical situations where bleeding can help a patient.

[‡‡‡]During the Second World War, the Army Blood Transfusion Service chose a vampire bat for its logo.

and intelligence. To learn all the skills needed to survive takes time and baby vampire bats are nursed for nine months, six months longer than similar-sized mammals. Living in small groups, they share information, food and adopt orphans. They show reciprocal altruism and are most generous to those who have helped them in the past.

Their body, as well as their behaviour, is also adapted to their dietary habits. Special cells in their nose that are sensitive to infra-red light help them detect the warmest, most blood-rich spots on a cow's body. Their teeth are huge, by bat standards, and don't have an enamel coating, so they are constantly being sharpened and can bite into the cow's hide without awakening it. They feed for around twenty minutes and have to keep the cow's veins open and the blood flowing. To do this, they secrete a mixture of compounds in their saliva that keeps blood flowing, free from clots, keeping the blood vessels open. They constantly lap at a wound with their grooved tongue, simultaneously catching the blood and delivering more anticoagulant into the wound from their saliva.[§§§]

Even having mastered the art of getting a good meal, vampire bats are restricted by the poor nutritional value of the blood. They have to drink up to 60% of their body weight in blood every night to get enough nutrition. Most of the blood they drink is water, so they have to excrete it almost as fast they drink to concentrate the much-needed protein. The low fat levels in blood means the bat can't lay in fat stores for lean times and must feed every couple of days to survive. Vampire bats share blood meals with others in their small colonies to keep the group alive and, unlike most bats, they can't hibernate or migrate long distances. If vampire bats are anything to go by, it perhaps explains why vampires spend the majority of their time doing very little but lying around in their coffins.

Vampires would have to have a very different metabolism to be able to survive on blood alone. Whatever adaptations have been made appears to work for them, and in some vampiric examples, blood can do more than just sustain the body, it can rejuvenate it. Fictional vampires like Dracula get younger and stronger as they

---

[§§§]The enzyme desmoteplase, from the saliva of the common vampire bat breaks up blood clots and has huge medical potential as a treatment for stroke victims.

drink more blood. While an all-blood diet is a very bad idea as far as human health is concerned, is there any truth in the suggestion that blood could make you young again?

4.5 THE FOUNTAIN OF YOUTH?

The idea of using blood to rejuvenate the old is not exclusive to vampires. The alleged real-life vampire Erzsebet Báthory was using blood as a beauty aid. If the stories are to be believed, and they might not be, she thought that young blood could transfer its youthful properties to her. She was not alone. Other historical figures have also tried to keep themselves young using blood.

The Romans drank the blood of dying gladiators to absorb their strength. In the fifteenth century, Pope Innocent VIII, when he was dying, had three youths brought to him and he either drank their blood, bathed in it or had it injected into his veins,[¶¶¶] because he wanted their youth and life. The Pope's plan didn't work and he died soon after. Emperor Jaijing in sixteenth-century China is said to have drunk the menstrual blood of young women to lengthen his life, and he kept a group of young virgins in his palace for his supply. He also experimented with other elixirs, many of which contained high levels of mercury and it was these that probably contributed to his death at the age of 60.

In more recent times, Alexander Bogdanov, a Russian physician, philosopher and science fiction writer, experimented with the rejuvenating properties of blood. In 1924, then aged 51, he began a series of blood transfusions, taking donations from his sister and several other volunteers, in the hope of achieving eternal life, or at least putting off the ageing process. After eleven transfusions, he was delighted to report that his eyesight had improved and his balding had been halted. His wife also benefited from donations that she claimed had not only cured her gout but made her feel great. In 1925, he set up the Institute for Haemotology and Blood Transfusions. Bogdanov's personal transfusions continued up until 1928 when he received his final and fatal donation.

One of his students was suffering from tuberculosis and malaria and Bogdanov believed that mixing their blood would

[¶¶¶]The details vary and the story only emerged after he died, perhaps conjured up by his enemies.

cure the student. The student survived, but Bogdanov suffered a severe reaction and died. Though he understood the basics of blood typing, he did not understand the role and importance of antigens and antibodies. With every transfusion, new antigens and antibodies had been introduced into Bogdanov's body. Each time his body had mounted an immune response to the foreign bodies invading his blood, adding to its library of antigens and developing a better, quicker response to the invaders. By the time of the last transfusion, Bogdanov's immune system was primed to launch a massive attack on the foreign blood flowing into his veins, and it killed him.

Today, with our much better understanding of antibodies and the human immune response to them, blood transfusions are very safe. As the number of blood transfusions has increased over the years, patterns have emerged. In an emergency situation, blood can save a life, but those who specialise in trauma care or who regularly carry out transfusions have noticed that younger blood is better, as is blood that is fresher. However, blood donations from all age groups are normally mixed and stored until they are needed.[||||] With these hints at some youthful property within blood, and with better understanding and safeguards in place to prevent disasters like that of Bogdanov, it might be tempting to try and use blood to halt the ageing process. Some scientists are doing just that, but slowly and systematically, and most often with animal models instead of humans.

In 1972, an experiment was conducted with rats. It was found that old rats given young blood lived four to five months longer than average, a 10–20% increase in life expectancy. In 2013, another study in mice showed that stem cells in older mice could behave like younger stem cells after receiving blood from younger animals. The experiments involved stitching two mice together, one older and one younger, to allow their blood to mix. The older mice developed shinier fur, a stronger heart and their injuries healed better. Clearly there was something about young blood that could help these rodents live longer and appear younger, but identifying what that something might be has proved tricky.

[||||] In the UK, red blood cells are stored for up to 42 days, but plasma can be cryogenically frozen and kept for up to a year.

The team who conducted the mouse study thought the rejuvenating effects might be due to a protein called growth differentiation factor 11 (GDF11). GDF11 is a protein that is expressed in higher levels in young animals and it is associated with several roles that promote good health and longevity. The identical protein is found in mice, rats and humans. The evidence was strong enough to support a human trial, which was subsequently set up to give plasma from young donors to eighteen Alzheimer's patients. By using plasma, the team could exclude any effects from blood cells, and though this narrowed the search for the fountain of youth, there are still many components within plasma that could be the key ingredient. It might also be several components working together that give the apparently miraculous effects. Trying to identify those components so they can be given on their own, rather than as a regular blood transfusion was important. Apart from the inconvenience of a treatment that involves regular transfusions, there are also problems with giving large amounts of whole blood to a patient. They can become overloaded with iron from the red blood cells, which is difficult to remove.

The experiment on the eighteen Alzheimer's patients was found to be safe, *i.e.* there were no obvious harmful effects, but there were only 'hints' that the patients got any better. Subsequent studies have shown that large amounts of GDF11 can be damaging rather than beneficial, but studies into other components of blood continue.

Blood components have also been trialled to combat Parkinson's disease and to boost brain function. Other researchers have suggested that it is toxins created by our own cells that age us and that dialysis could be used to remove them from older blood. However, some aren't waiting for scientists to home in on the magic ingredient.

If you are over 35 and have $8000 to spare, you can take part in the Ambrosia trial, where you will be injected with two litres of plasma taken from young blood donors. The aim of the trial is not to cure any particular disease but to relieve 'the general malaise of being old'. The cosmetic industry has also not been slow to jump on the blood bandwagon. Vampire facials are available to the rich and vain, though their benefits are far from proven. This time, it is plasma from your own blood that is re-injected to plump up the face and remove wrinkles.

None of these trials or treatments has shown the dramatic rejuvenating effects experienced by Count Dracula. He went from a grey-haired old man to a strong young man with jet black hair in just a few weeks of feasting on young blood. But there have certainly been some intriguing results. Should these blood therapies take off, there might be some interesting vampiric parallels to be made if older people are to benefit from taking the blood of the young. For a start, will there be enough to go around?

As far as medical and scientific research is concerned, many devices and processes have been devised to make donated blood stretch as far as possible. Improved methods of storage means blood can be kept for longer and by separating out the different components of blood, one donation can be used for several different purposes. Vampires drinking donated blood, as they do in the TV series *Angel*, would be safer for the public in one respect, but adds another drain on a precious resource. Synthetic blood could potentially solve a lot of problems, from shortages of real blood donations, to preventing vampire attacks.

4.6 SYNTHETIC BLOOD

From a nutritional point of view, one type of blood is much the same as another and so some vampires in folklore and fiction have resorted to feeding on other animals to keep them sated. Arnold Paole was said to have drunk blood from cows, and Louis in *Interview with the Vampire* fed on rats. In the TV series *True Blood*, a scientist came up with an even better solution, synthetic blood. It might not be that difficult to create an artificial blood that had the same nutritional content as natural blood, though it might not look or taste as good as the real thing. However, blood would appear to give vampires much more than just a good meal. These extra qualities may not prove so easy to mimic, as has been shown in real-life attempts to make artificial blood or blood substitutes.

Take humans, for example. Blood does so much more than feed us, it keeps us alive and healthy. Despite many attempts, it has proved difficult to create a synthetic version that can carry out blood's many functions as successfully as the real thing.****

****In the seventeenth century Christopher Wren proposed a blood substitute made from ale, wine and opium. It wouldn't have performed any of blood's normal functions but it probably made you feel too good to care.

However, in an emergency situation just being able to keep someone alive long enough for treatment is an advantage. Though whole, fresh blood is undoubtedly the best option, substitutes could potentially give someone valuable extra minutes in which medical professionals could work to save a life.

In the case of severe blood loss, the main aim is to replace the lost volume of blood and to keep a good supply of oxygen going through the body. Because whole blood is not always readily available in civilian situations, many trauma teams use mixtures of equal parts of plasma, platelets and red blood cells. The plasma replaces the lost volume, the platelets help to stop further bleeding, and the red blood cells help carry oxygen.

If the goal is simply to replace the lost volume of blood, plasma substitutes should be the easiest to mimic artificially. Crystalloid and colloid transfusions were commonly used, up until the 1990s, to take the place of plasma and resuscitate patients in emergency situations, until blood products became available. These were solutions of the key ingredients normally found in blood plasma, such as salts and albumin. The crystalloid formulation was not very successful, as the water in the solution tended to be drawn away from the blood and absorbed into the body, causing problems such as pulmonary oedema. The colloid preparation was better, but more expensive, and still not as good as real plasma.

Red blood cell substitutes are more difficult than plasma substitutes. Some have sought to simplify the problem further by focussing on the haemoglobin component of red blood cells. However, if haemoglobin is free to roam in the veins and arteries it scavenges nitric oxide, the chemical that signals for blood vessels to dilate. The consequence is that when haemoglobin escapes the confines of the red blood cells, blood vessels constrict, blood flow is restricted, and this can lead to strokes and heart attacks. Modifications have been tried where multiple haemoglobins are tethered together or immobilised on an inert polymer.

Other alternatives have been tried that are chemically completely different to haemoglobin, but are still able to absorb and release oxygen in the right places. Perfluorocarbon (PFC) based products can dissolve oxygen into molecular cavities and release it again in areas of the body that are oxygen deprived. The molecules are inert and after infusion, the molecules vaporise and are exhaled. PFC systems have been used to provide oxygen to premature babies, but the system is not a substitute for whole blood. Of the many synthetic alternatives that have been tried, none have yet proved completely successful.

Whole red blood cells would avoid the problems of free-roaming haemoglobin, and there are ways of growing human cells in laboratories. But red blood cells cannot be grown in a lab the way extra skin cells can be grown for skin grafts, for example. Red blood cells don't have a nucleus, so they don't have the DNA instructions to reproduce on their own, but some researchers have shown ways to use stem cells to make red blood cells. This is a potentially much less risky method than either donations or synthetic blood. The red blood cells would be produced in sterile conditions, reducing the risk of transmitted disease that has blighted volunteered blood donation schemes in the past.[††††] The red blood cells could also be produced as type O, without the antigens on their surface that cause rejection problems. The lack of nucleus can also be an advantage, as these red blood cells can't turn cancerous. Though there is huge promise from this line of research there is still a long way to go before laboratory-grown red blood cells find their way into hospital trauma departments.

Alternatives to platelet cells are even more complex. Getting artificial platelets to clump at just the right place at the right time is tricky. Getting it wrong can mean thrombosis or uncontrolled bleeding. Accept no substitutes, for the time being anyway. The components of blood work together in a wonderfully elaborate inter-dependent system that carries out a multitude of roles in the human body with remarkable efficiency. It is no wonder vampires crave it.

[††††]A scandal in the UK saw thousands of patients with haemophilia infected with hepatitis C and HIV through contaminated blood donations. CJD, the human form of mad cow disease that affected the UK also means UK blood donations are not always accepted in other countries.

CHAPTER 5

Sunlight

> 'Vampires generally arrive at night, in carriages drawn by two black horses. Our Vampire arrived by the commonplace means of the railway train, and in the afternoon.'
>
> Count Stenbock, *The True Story of a Vampire*

Every evening, Robert Neville had to barricade himself inside his home for the night. His windows and doors would be barred and smeared with garlic in an effort to keep out the undead. Outside, swarms of revenants – some of them his former friends and neighbours – would gather to entice, taunt and threaten him in their attempts to make him venture out so they could drink his blood. In the morning, before sunrise, the street would empty and the vampires would slink away to hide in the shadows, out of the sunlight. Robert could then safely emerge.

The vampire-like undead in Richard Matheson's 1954 novel, *I Am Legend*, have many of the characteristics normally associated with this type of reanimated corpse – they drink blood, they have pale skin and they only come out at night. They are not simply nocturnal, but fall into a deep trance-like state from which it is impossible to rouse them before the sun goes down. It was therefore safe for Robert to search out as many of these creatures as he could find during the day and either stake them or drag them into the sunlight to destroy them.

Vampirology: The Science of Horror's Most Famous Fiend
By Kathryn Harkup

Published by the Royal Society of Chemistry, www.rsc.org

Vampires seem to have a very strange relationship with light, and sunlight in particular. They may be well known for their aversion to the sun's rays, but the characteristic is not unique to them. There are plenty of humans who prefer to avoid the sunshine, or daylight entirely, and stay indoors, but there can be greater consequences for vampires if they choose to sunbathe. In the past, folklore vampires were usually, but not always, nocturnal creatures, but out of habit rather than compulsion to flee daylight.† However, modern vampires are generally believed to have more than a strong dislike of sunlight, it is an outright fear that the sun's rays will destroy them instantaneously in a ball of flames. The same vampires who fear being burnt to a pile of ash in the sun's rays are often shown sitting under electric lights without any discomfort, and in some fictional vampire stories moonlight has the opposite effect on them, restoring them to full health after they have been 'killed'. When it comes to light, why are vampires so very different from us?

5.1 THE LIGHT FANTASTIC

Light is a form of energy and its spectrum ranges from long-wavelength, low-energy radio waves, to short-wavelength, high-energy gamma rays. Receptors in our eyes are sensitive to a narrow range of wavelengths that we call visible light, a band of wavelengths that sits roughly in the middle of the light spectrum.‡ And although we can't see all the wavelengths along the spectrum, we can perceive the effects of some of them; for example, the warmth we feel from infrared (IR) light and the tanning effect of ultraviolet (UV) light.

Different materials absorb different wavelengths of light, making some materials opaque and others transparent, but all forms of light along the spectrum obey the same basic rules of reflection and absorption. Our body is mostly opaque to visible light, meaning the light is absorbed and we cast a shadow, while some of the light is reflected off our body and into our eyes or a camera, meaning we can see people or take photos of them.

†Vampires in Russia and Poland were said to prowl around from midday to midnight and Dracula was spotted in London in broad daylight.

‡Other animals can have receptors tuned to different wavelengths of light, giving them a very different view of the world.

Vampires would appear to be an exception to the fundamental laws of physics surrounding how light behaves, as they cast neither shadows, reflections in the mirror, nor appear in photographs. This is extraordinarily difficult, if not impossible, to explain scientifically. If vampires really are the solid bodies they are generally held to be, and not incorporeal spirits, light should interact with them in the same way that light is reflected, refracted or absorbed by any other solid substance. If vampires can see themselves, for example, by holding their own hand in front of their face, they should also be able to see their reflection in a mirror. If *you* can see vampires, you should also be able to take a photo of them. The reasoning behind these strange vampiric characteristics probably has more to do with ancient beliefs that reflections and shadows were related to your soul, and being dead, vampires had no soul.

But, what about different forms of light, such as sunlight *versus* electric light or moonlight? Could these have a different effect on vampires? The light emitted by the sun covers most of the spectrum, from radio waves to X-rays, but the most intense output is in the visible region, with quite a bit of IR and UV thrown in. The Earth's atmosphere absorbs a lot of the harmful UV before it reaches us on the ground (more on this later). If vampires are particularly sensitive to visible, IR or UV light, then perhaps sunlight is too strong for them.

Light from candles or light bulbs is much less intense than direct sunlight. Though a room may appear very brightly lit, the pupils of our eyes adjust to different light levels, so that we aren't dazzled in bright light or struggling to see in low light. Our eyes constantly deceive us as to the true level of light, which can be measured independently and accurately using a light meter. Artificial lights can also emit different wavelengths of light, compared to the sun, because they are not nearly as hot and are made of specific materials that radiate specific wavelengths. Perhaps the low intensity, or lack of certain wavelengths of light means vampires are not adversely affected by artificial light in the way they apparently are by sunlight.

The other light source that features in vampire fiction, and particularly in Polidori's *The Vampyre* and Rymer's *Varney the Vampire*, is moonlight. In both these stories, moonlight is not just preferred by the vampires, it can fully restore them after they have been 'killed'. Moonlight has strong associations with the supernatural, from werewolves to witchcraft, so is there something special about this light?

The light that shines on us from the moon is reflected sunlight, and a tiny amount of reflected light at that.[§] Moonlight, depending on the phase of the moon, ranges from 0.1 to 0.3 Lux (by comparison, direct sunlight ranges from 32 000 to 100 000 Lux). And because it is reflected light from the sun, it has an almost identical composition, or spectrum, in terms of wavelengths. Moonlight contains no special moon rays that could affect vampires any differently than sunlight, though its much lower intensity might explain why vampires prefer it if intense light harms them. But, having said all that, there are still ways that light could affect a vampire differently to a human. Humans depend on light in many ways, not just in helping us to see the world around us.

5.2 FROM DUSK 'TIL DAWN

Sleep is essential for human health. Lack of sleep doesn't just make us grumpy and tired, prolonged experiences of poor sleep can have a seriously detrimental affect on our health, from obesity to heart disease, and can shorten life expectancy. In extreme cases of sleep deprivation, people experience poor coordination, hallucinations and seizures.

While it is clear that humans need sleep, what is less clear is why. The production of certain hormones, such as growth hormone, peaks during sleep. It is also an important time for cell reproduction, allowing us to grow and repair everyday wear and tear. One theory suggests that sleep is an opportunity for the brain to cleanse itself of toxins that build up during wakefulness, like taking a nightly shower. Another theory is that we sleep to save energy. During sleep our body temperature drops slightly and our metabolic rate slows. The metabolism theory is backed up by comparing the metabolic rate and average sleep of a variety of animals. Generally speaking, smaller animals have a higher metabolic rate and sleep longer than bigger animals. And, though there are some notable exceptions to the rule, it might

[§]There is also a very small contribution to the moon's rays from reflected starlight and light from Earth.

explain why vampires, with their low-energy blood-based diet, need to sleep so long in their coffins.

How the body regulates these periods of sleep and wakefulness depends on intricate and interconnected biological processes. We have a powerful biological clock governing cycles of sleep and wakefulness in our bodies. These cycles are known as circadian rhythms from the Latin *circa*, meaning around, and *diem,* meaning day. But there is much more going on in the body than simply feeling sleepy or more awake at different times of the day. Almost every cell in our body contains its own circadian clock governing when certain process should be carried out. Molecules produced at one point in the cycle build up until they reach concentrations that trigger other processes in the body. Take, for example, adenosine, the molecule involved in energy processes in cells that we met in Chapter 3. Adenosine builds up in the brain when we are awake until it reaches levels that trigger feelings of sleepiness.¶

There are many other chemical peaks and troughs operating to keep the body on a roughly 24-hour cycle, but overall control is maintained by a tiny region of the brain called the suprachiasmatic nucleus (SCN). The SCN continuously sends and receives messages to regulate the cycles going on in the body. To avoid the internal body clock drifting out of a regular 24-hour pattern, the SCN synchronises itself with the outside world. Unsurprisingly, one of the main external cues for the internal body clock is light. The SCN, part of the hypothalamus, is situated deep within the brain and just above the optic chasm, along which signals are transmitted from the eye to the visual region of the brain.

Low levels of light entering our eyes triggers the pineal gland to produce melatonin. This hormone is released in many different animals but it causes different things to happen. Though it is sometimes referred to as the 'sleep hormone', melatonin is better described as the 'darkness hormone', because it tells an animal what to do when it gets dark. For humans, increasing melatonin levels tell us to sleep, but in nocturnal animals, and perhaps vampires, it has the opposite effect. Spending large amounts of time in a completely dark grave, coffin or crypt must mean a vampire's melatonin levels are very high by the time the sun sets and they get up to go about their nightly business. Experiments in

---

¶Caffeine blocks the action of adenosine, making you feel more alert.

humans have shown that without external light cues or clocks, our cycle runs a little longer than 24 hours and over days or weeks we can easily become very out of sync with the world around us.‖ Vampires will have few, if any, external light sources to calibrate their nightly and daily cycles, so either their endogenous or internal biological circadian rhythm is almost exactly 24 hours, or they have very good alarm clocks to wake them after the sun has gone down.

Nocturnal behaviour has other consequences, because light doesn't just regulate our biological clocks. Avoiding daylight completely may present another problem for vampires. If they are anything like humans, and they may not be, they will need vitamin D to maintain a healthy existence. This vitamin helps regulate calcium and phosphate in our body, which is essential for healthy bones and teeth as well as muscles, but its roles extend far beyond that. It helps cells proliferate, so the body can grow and repair itself, as well as affecting the immune system. Low levels of vitamin D have been linked to increased severity of infections such as HIV and tuberculosis. And there are many more jobs it helps with – almost every cell in the human body has receptors for vitamin D.

Humans convert 7-dehydrocholesterol, a side product of a cholesterol-forming system, into vitamin D. Production of 7-dehydrocholesterol isn't the problem; it is abundant in the skin, but the process to convert it into vitamin D requires UV light. No animal on earth can carry out the conversion without help from the sun. If not enough sunlight reaches the skin, then vitamin D has to be obtained directly from the animal's diet. Fortunately, drinking four litres of blood a day (as calculated in Chapter 4) would provide more than enough vitamin D courtesy of the vampire's victims. However, most human diets simply don't contain enough vitamin D and we are reliant on sun exposure.** This can pose a problem for some people.

Melanin, the dark pigment in skin, protects people from the damaging effects of the sun and for those living near the equator it is essential to prevent skin cancer. It also prevents the sun's rays from reaching the parts of the skin that produce vitamin D. For those living in sunny climes near the equator this is rarely a problem as there is so much sunlight. However, those who cover

‖Those who are totally blind often need clocks to maintain a regular sleep/wake cycle.
**Oily fish and fortified dairy products are good sources of vitamin D.

much of their skin for religious reasons, or people who live further north, may not get enough sun exposure.[††]

People with pale skin living in more northern regions, where the sun doesn't shine as often and the amount of sunlight reaching the ground is less intense, have the opposite problem. A reduced amount of melanin in the skin allows more sunlight to reach the parts of the skin where 7-dehydrocholesterol can be converted to vitamin D. For pale skinned people in higher latitudes, between ten and thirty minutes of sun exposure to the forearms and legs a day is plenty to ensure enough vitamin D is produced.[‡‡] Problems occur for pale-skinned sun-worshippers, and Europeans who have relocated to sunnier climes, who spend longer in the sun and fail to take adequate protection. They run the serious risk of too much exposure to the sun, leading to sunburn and potentially skin cancers. Humans can learn from vampires – too much sun is definitely a bad thing.

It is the UV component of sunlight that causes all the problems: on the one hand, it makes vitamin D but on the other hand it can damage DNA. There are three bands of UV light emitted by the sun: from least to most damaging, they are UVA, UVB and UVC. The UVC can be ignored because the Earth's atmosphere absorbs it all before it reaches the ground. Much of the UVB is also absorbed but some still shines down on us. It is the UVB range that allows vitamin D to be produced, but it is also powerful enough to pull apart the DNA in our skin cells. We have repair mechanisms to correct the damage, but it is not perfect; mutations caused by UVB can accumulate and lead to cancer. UVA does not have enough energy to damage DNA directly, but it can oxidise the melanin in the skin, which can then go on to cause DNA damage.[§§]

Anyone who spends too much time in the sun runs the risk of damage to their skin but for some the risk is much, much higher. Some medical conditions expose sufferers to much more sun damage, and in extreme cases people with these conditions avoid the sun completely. Could these medical conditions offer an explanation for the origins of the vampire's sun-avoiding characteristic?

[††]Vitamin D supplements can help correct deficiencies.

[‡‡]However, there are huge variables depending on skin tone, clothing, cloud cover, air pollution, time of day and season.

[§§]Always check your choice of sunscreen protects you against both UVB and UVA.

5.3 XERODERMA PIGMENTOSUM

Xeroderma pigmentosum is a genetic condition that affects around one in every one million of the human population in Europe and the USA. The genetic mutation means the repair system that fixes the damage to our DNA caused by UV light is completely absent. As a result, skin cancers occur easily and their development accelerates rapidly. There is no treatment available other than to avoid sunlight, and it forces its sufferers into a nocturnal lifestyle. At Camp Sundown in the State of New York, the day is reversed so children can play outside during the hours of darkness. Without such provision, as is common for those living with the condition in developing countries, survival beyond the teenage years is rare. Xeroderma pigmentosum sufferers are often called 'moonlight children', but other than an understandable aversion to sunlight, they have nothing in common with vampires. Other genetic conditions, however, have more coincidences with vampire characteristics and have been touted as possible explanations for the origins of vampire stories.

5.4 ALBINISM

People with albinism are more adversely affected by sunlight than others. Albinism is a genetic condition that means pigmentation in the skin, hair and eyes does not form. The result is exceptionally pale skin and hair and eyes that can appear red, simply because the lack of pigment in the iris means the red-coloured interior of the eye is more visible. As a consequence of the lack of pigmentation, these people have minimal protection from the damaging effects of sunlight on the skin. There can also be problems with vision, including difficulty adjusting to bright lights such as strong sunlight. Normally the eye compensates for different levels of light by widening or constricting the pupil by moving muscles within the iris. Though the iris will still alter in an individual with albinism and can constrict the size of the pupil, the iris itself can let light through to the interior.

In Tanzania, the number of people with this genetic variation is particularly high, and they are especially vulnerable to skin cancer living so close to the equator. For some, however, cancer is almost a secondary consideration to the social stigma of the condition because they are often persecuted for their appearance. Special schools with high walls protect their pupils, not from the sun, but from people outside who believe albinism brings with it special magical and curative properties. For example, some think the children are demonic spirits, others believe the crushed limbs of children with albinism can cure any and every ailment, and consequently they fetch a high price. None of these ridiculous assumptions made about albinism has any basis in fact.

People with albinism may have a striking appearance. Their pale skin and reddish eyes coincide with ideas of what literary and film vampires should look like. They also have a much greater need to cover up in strong sunlight, and many wear tinted glasses to protect their eyes from the glare of the sun. These very superficial similarities to a few characteristics of vampirism, however, do not make them vampires.

The idea of the pale vampire only emerged in the nineteenth century. It was a reflection of the fashions of the day and associated with people like Lord Byron, who was considered to be irresistibly handsome at the time (in part) because of his clear pale skin. Before Polidori's vampire appeared in 1819, vampires were almost always described as having a ruddy complexion. The image of the pale vampire was perpetuated by early cinema and has now become a firmly established characteristic in popular culture, but this will be discussed more in later chapters. Red eyes were introduced with *Dracula* and have been repeated in many of the film versions of the novel. In the past, people with albinism would not have been considered vampires and they are not the origin of vampire legends.

Albinism is far from the only medical theory put forward as an explanation for vampirism, but it is probably the worst in terms of matching symptoms to vampire characteristics. A more convincing theory was published in the *Journal of the Royal Society of Medicine* in 1997 based on pellagra, a condition brought about by dietary deficiencies.

5.5 PELLAGRA

Pellagra is often identified clinically by the 'four Ds': dermatitis, dementia, diarrhoea and death. The details of these symptoms have some remarkable similarities to characteristics that are commonly associated with vampires, in particular their sensitivity to sunlight. Pellagra was known as spring sickness in some parts of the world because symptoms started to appear as the days got longer and warmer and people spent more time outside. It would be worse during the summer and fade again in the winter months.

Pellagrins, those suffering from pellagra, are hypersensitive to sunlight, which causes the first of the four Ds, dermatitis. Areas of skin that are most exposed to sunlight – such as the face and hands – become red, the skin thickens and becomes scaly. There is a characteristic 'broad collar' rash in a circle at the base of the neck, known as a 'casal collar'. The appearance of the rash is followed by inflammation and oedema, and eventually depigmentation. The person is left with patches of pale, shiny skin alongside patches of darkened, rough, scaly skin. As the condition progresses, the skin becomes paper-thin and has a parchment-like appearance.

The red rashes on the hands, feet and collar area of the skin – characteristic signs of pellagra – were first described by Gaspar Casal, a Catalan physician, in 1735. His observations were published in 1762 in a book called *Historia Natural y Medicina del Principado de Asturias* (*Natural and Medical History of the Principality of Asturias*) and, therefore, the condition was known for a while as 'Asturian leprosy'. The name pellagra, perhaps derived from the Latin *pellis* for skin, was given to the condition in Italy, where it affected over 100 000 people in the 1880s. In the early twentieth century the same disease appeared in the southern states of America and claimed the lives of more than a thousand people. The pockets of outbreaks across Europe and America coincided with regions that depended heavily on corn, or maize, for food and it was initially thought that some germ or bacteria was affecting the crops. The problem was indeed the corn, but not because of some germ or parasite it was carrying.

Corn is native to the Americas but was introduced into Europe with the return of Columbus and other explorers to the New World at the end of the fifteenth century. It had begun to replace rye and wheat as a staple crop because it provided more calories per acre. What it didn't provide was sufficient niacin and tryptophan. Both these compounds are in the plant, but the niacin is largely inaccessible because it is bound up in hemicellulose that can't be easily broken down in our digestive system. The result is that maybe only 10% of the niacin and tryptophan is absorbed into our bodies through our diet unless the corn is first treated, or nixtamalised (soaking and cooking in an alkali solution, usually limewater), to release it. This practice was common among native Americans who relied on corn for much of their diet, but was not normally carried out by Europeans.

Tryptophan is one of the amino acids that are essential for humans. It is a precursor for the melatonin hormone associated with regulating our sleep/wake cycles, but it does much, much more. In the body, tryptophan can be converted into niacin, also known as vitamin B3 or nicotinic acid. Niacin can be obtained from a wide range of different foods, including meat, fish and eggs, but in poor areas, where food options were limited, pellagra took its greatest toll. In the body, nicotinic acid is converted into nicotinamide adenine dinucleotide (NAD) and can then also have a phosphate added to make nicotinamide adenine dinucleotide phosphate (NADP). NAD is used by over 400 enzymes to catalyse reactions. NADP is used as a coenzyme by many more enzymes. Between them, NAD and NADP are used in almost every metabolic process in the body. Disruptions to these metabolic processes are what lead to the 'four Ds' that are used to diagnose pellagra.

Without sufficient niacin, metabolic processes in nerve cells degenerate. Pellagrins may have trouble sleeping. There is a restlessness, and an aversion to bright lights and strong smells. They can become aggressive without provocation, anxious and experience depression. The symptoms can be severe enough for people to be admitted to a psychiatric hospital and the condition can eventually lead to the development of dementia.

The third D, diarrhoea, can lead to severe weight loss. Pellagrins may shun food, perhaps because of the lesions in the mucous membranes that develop with the condition, leading them to develop anorexia.[¶¶] Gastrointestinal bleeding can also led to anaemia (see Chapter 4). Taken together, these signs are very similar to those shown by the vampire's many human victims. Pellagrins are also said to have a 'foul mouth' because of their pungent breath, an affliction also noted to have cursed Count Dracula. There is also an excessive redness to the tongue

[¶¶]An eating disorder characterised by a markedly reduced appetite or a complete aversion to consuming food.

that could, perhaps at a stretch, be mistaken for the traces of a blood meal. Finally, there is death, the last of the four Ds. Death in untreated pellagra victims usually occurs after four or five years, but can appear to be more sudden if symptoms are initially mild, but then are followed by a sudden decline in health.

There are many points in favour of the pellagra-vampire theory: the peeling skin and redness coincides with folklore descriptions of vampires and even the pale parchment-like skin of the later stages of the disease are similar to descriptions of the Count in *Dracula*. The aversion or avoidance of the sun to relieve symptoms does coincide with the modern vampire's fears about sunlight. The insomnia, aggression, refusal of food and possible aversion to strong smells, such as garlic, also support the theory as an explanation for the origin of vampire myths in Europe, but there are also significant points counting against it.

Accounts of the disease make no mention of drinking blood, though it might actually be a good source of tryptophan in the amounts vampires are expected to consume. The time scales for its progression are also not a good match with either folklore or fictional descriptions of vampirism. Seasonal outbreaks of pellagra, which started in spring and stretched through the summer to autumn, do not match well with folklore vampire epidemics that often occurred in winter or early spring and usually lasted a few weeks. Nor does it match the experience of a vampire's victims, which in fiction may be a slow drawn-out decline before they die, but it is usually described as taking days or weeks, rather than years. In folklore, the timespan is typically much shorter; those bothered by a vampire usually died within 24 hours of their attack. And what of the vampires themselves? Four to five years of suffering for a pellagrin is also too short for the modern vampires who are blessed, or cursed, with eternal life. Folklore vampires usually only enjoyed a few weeks of an undead existence before they were tracked down and staked. Perhaps without this fatal intervention they would have survived longer, but I guess we will never know if the natural life (or death) span of a folkloric vampire stretched to several years.

Would people in the eighteenth century have seen people with pellagra and thought they were vampires? Probably not. But as scientific explanations for the origins of vampire myths go, pellagra is not the worst theory and certainly ticks significantly more boxes than some others that have been put forward. However, it has not received nearly as much publicity as another theory of vampirism that focussed particularly on their extreme reaction to sunlight.

5.6 PORPHYRIA

Porphyria is a rare genetic condition that hinders the normal biosynthesis of haemoglobin (the oxygen-carrying component of red blood cells we met in the previous chapter). To understand why porphyria might be discussed in the same breath as vampires we need to first look at haemoglobin and how the haem, the most important part where the oxygen binds, is made.

One oxygen molecule bonds to a single atom of iron (Fe) held at the centre of haem, as shown in Figure 5.1. The arrangement of carbon and nitrogen atoms in hexagons and pentagons around the central iron atom is called a porphyrin. The porphyrin is like an elaborate collar, or ruff, that keeps the iron in place and modifies its behaviour. It is practical but also decorative as the pattern of single and double bonds (the double and single lines) also means porphyrins are strongly coloured. It is the porphyrin unit together with the iron atom, a tiny fraction of a much larger protein that gives blood its intense red colour. The subtle changes that occur when an oxygen bonds to the iron slightly modifies the other bonds in the structure and we therefore see an appreciable colour change between dull red/brown deoxygenated blood and bright red oxygenated blood.

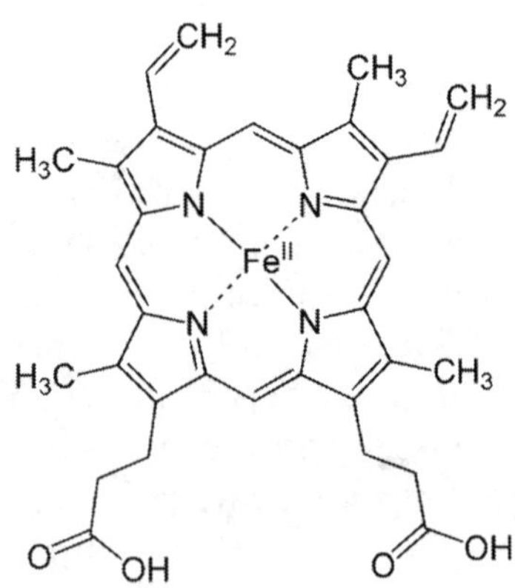

Figure 5.1 The structure of haem.

Each molecule of haemoglobin contains four iron-centred haem units and each red blood cell contains millions of haemoglobin molecules. All of these millions of molecules have to be made from the material available, *i.e.* what we eat. Food gets broken down into smaller and smaller pieces as it travels through our digestive system until these fragments are small enough to be absorbed into our body proper. These fragments, small molecules, are used as the building blocks that construct the millions of different chemical components that keep our body functioning normally. Animal and plant proteins that we eat are broken up into their constituent amino acids that can then be restrung in a new order to make the protein part of haemoglobin. However, the all-important haem unit has to be made almost from scratch.

Making haem is remarkably efficient in terms of raw materials. Each one is made from only three basic ingredients; glycine (an amino acid), succinyl coenzymeA (a by-product of the citric acid cycle), and an iron atom. By using repeat units, and a few chemical modifications along the way, a complete haem can be made in just eight steps before it is attached to the rest of the protein to make haemoglobin (see Scheme 5.1).

Each of the eight haem-making steps is controlled by a specific enzyme, like a production line with robots carrying out specific tasks at each point along the way. Each enzyme adds a new unit, or modifies the structure, ready to go on to the next stage. It is a very efficient process where, normally, none of the haem precursors hang around in the body for any length of time because they are rapidly converted to the next stage along the production line.

A genetic mutation in any one of these eight enzymes can result in the whole process slowing down by creating bottlenecks in the production line and a build-up of haem precursors. Mutations in the enzymes involved in any of these steps results in diseases collectively known as porphyria. There are several different types of porphyria, each with different symptoms depending on how the enzyme has been mutated and which of the haem precursors is forming a backlog in the system.

Having the genetic mutation for one of the forms or porphyria does not necessarily mean an individual will have symptoms. Many will spend their entire lives without suffering the ill effects of the condition. For some porphyria patients, the symptoms

Glycine + Succinyl-CoA

↓

5 Aminolevulinic acid

↓

Porphobilinogen

↓

Hydroxymethyl bilane (→ Uroporphyrinogen I)

↓ **congenital erythropoietic porphyria**

Uroporphyrinogen III

↓ **porphyria cutanea tarda**

Coproporphyrinogen III

↓

Protoporphyrin III

↓

Protoporphyrinogen IX

↓

Haem

Scheme 5.1

appear later in life, and for many it is not a continuous affliction. Episodes of illness can be triggered by stress, certain prescription drugs or changes in hormone levels.

In the case of porphyria cutanea tarda (PCT), the most common from of porphyria, it is the fifth step in the process that gets held up. The slowing down of the conversion of one haem precursor into the next results in a backlog of uroporphyrinogen III. The body has no mechanism for getting rid of this haem precursor as it shouldn't be there in the first place, and so levels build up in the body.

Uroporphyrinogen III, like haem, is also dark in colour. The more intensely coloured or darker an object is, the more light it absorbs. During episodes of PCT, some excess uroporphyrinogen III will be expelled in urine and, after standing in sunlight for a while, the urine will turn a dark red or purple colour, hence the name 'porphyria' from the Greek *πορφύρα,* meaning purple.

A build-up of the same molecule in the skin also increases its sensitivity to sunlight, leading to exposed skin becoming fragile and blistering. These wounds can be slow to heal and can leave small scars.

A much rarer form of porphyria produces a much greater sensitivity to sunlight. In people with congenital erythropoietic porphyria (CEP), or Gunther's disease, it is a failure of the enzyme involved in the fourth step of haem synthesis. More than 35 different mutations in the gene that makes the enzyme in question have been identified. The different changes to the enzyme results in varying efficiency and a wide range of severity of symptoms depending on how much uroporphyrinogen I builds up in the body. Uroporphyrinogen I is a reddish-brown molecule and very similar to uroporphyrinogen III, but with more dramatic impact on the body.

In addition to dark urine, the presence of this haem precursor is particularly noticeable in the teeth because of the dark brown or red appearance it can give them. It can also cause the gums to recede, making the teeth look longer. But, most significantly, uroporphyrinogen I can build up in the skin, making it extremely photosensitive. There can be chronic blistering on areas of skin exposed to sunlight and they will be slow to heal. Untreated blisters can become infected and lead to disfiguring scars and, in the very worst cases, loss of fingers and facial features such as the nose and lips. This form of porphyria is exceptionally rare and only a fraction of those affected by it show the worst symptoms.

Not all forms of porphyria cause hypersensitivity to light. For example, porphyria is thought to have caused the periods of illness that afflicted King George III. His urine turned purple and he was extremely unwell, but he had no heightened sensitivity to sunlight. In the case of light-sensitive porphyrias, protection from the sun is difficult. Suntan lotion will not block the light that is causing the damage, but the skin can be protected by clothing and by staying out of the sun as much as possible. One treatment for CEP also has potential links to vampirism. In extreme cases, phlebotomy (bleeding), can be used to remove the toxic uroporphyrinogen I, followed by transfusion of donated blood full of healthy red blood cells packed with haemoglobin.

In 1985, Dr David Dolphin proposed porphyria as a possible explanation for vampires in an address to the American

Association for the Advancement of Science. Dolphin suggested that isolated populations in Central and Eastern Europe may have increased the chances of inbreeding, decreasing the genetic diversity and increasing the chances of someone being born with a genetic disorder, such as porphyria.

To back up his porphyria-vampire theory he suggested that because those affected would be desperately short of haem, they would be driven to seek blood. In the days before blood could be transfused into the veins, these people might have been compelled to try and drink human blood to relieve their agonising symptoms. He also noted a compound found in garlic, diallyl sulphide, promotes changes in the activity of liver enzyme p450, and that could exacerbate the symptoms of porphyria. He further claimed there were anecdotes that those who suffer from the condition avoided garlic.

On the face of it, this may seem like a convincing theory – red teeth, sunlight burning the skin, a compulsion to drink blood and an aversion to garlic – but it fails on almost every point. People with porphyria are people with porphyria. They are not vampires and have never been thought to be vampires in the past. A good scientist knows that anecdotes are not data and, without any proof to back it up, the garlic aspect of the theory should be dismissed out of hand.

Almost as soon as Dolphin had announced his theory, it was pointed out that porphyria sufferers were not compelled to drink human blood. The protein part of haemoglobin would be digested away, but it is true that some of the haem component of blood can survive the digestive process in humans to be absorbed into the blood stream from the intestines. However, there is no evidence that, even though the haem could circulate in the blood, it could be directed to the bone marrow and incorporated in the production of new haemoglobin and red blood cells. You would also have to drink a lot of blood to get enough haem into your system, and the problems associated with that were outlined in the previous chapter.

The final nail in the coffin for the blood-drinking theory comes from the history of the disease. Though it was first described by Hippocrates in 370 BC, the underlying biological cause wasn't established until 1871 by Felix-Hoppe Seyler. Why would a porphyria sufferer in the eighteenth century drink blood if they didn't know that red blood cells and haemoglobin was what they needed?

After Dolphin's announcement, folklorists were also quick to point out that vampires in folklore tradition showed no aversion to sunlight. This was the biggest blow to the theory. The emphasis on some forms of porphyria making an individual exquisitely sensitive to sunlight was its biggest downfall and showed that, though Dolphin may have researched porphyria, he had made very scant investigation into vampire traditions. His theory was shot down in flames very quickly, but not before it had gained huge publicity, despite its obvious and many flaws, and did a great disservice to people with porphyria along the way.

David Dolphin was not the only one to make a link between porphyria and supernatural creatures, as others have used it as an explanation for werewolves. This was mainly based on the fact that the damaged skin of porphyria patients, after it heals, often shows increased hair growth, and the areas of skin most affected are usually the face and hands, because they are most exposed to sunlight. However, the excess hairs that grow are very fine. Werewolves in folklore and popular cinematic portrayals are unmistakably either fully transformed into wolves, or show growth of extremely thick, long hairs all over the body. In the past, individual porphyria sufferers, on top of suffering the effects of a horrendous disease, have also been labelled werewolves. However, the belief in werewolves was already in existence, these people did not create the myth.

The medical conditions described above clearly illustrate the potentially damaging effects of sunlight, but it's not all bad news. If harmful creatures like vampires can be destroyed by sunlight, why not other harmful creatures like bacteria? Some have seen ways to subvert the harmful effects of sunlight and put light to positive use. One of those people is Dr Eric Skaa, who used vampires as a metaphor for a medical condition and its potential treatment. In July 2017, he tweeted: 'If *S. aureus* is going to drink our blood like a vampire, let's kill it with sunlight.' He was talking about using light to treat a bacterial infection, which is just one of the many diseases that can be potentially treated by photodynamic (or light) therapy.

5.7 PHOTODYNAMIC THERAPY

The '*S. aureus*' that Dr Skaar was referring to in his tweet, is *Staphylococcus aureus*, a bacteria that lives on the skin of roughly one third of all humans, usually without causing any problems. However, in some people whose skin is compromised, such as those who suffer from eczema, it can be a different story. Where the skin is broken, *S. aureus* can make its way through the cracks and release toxins, such as exfoliatin, that can contribute to the pain and persistence of the inflammation of eczema. In children under the age of five, the reaction to the toxins can be extreme, causing staphylococcal scalded skin syndrome. The top layer of skin can peel off leaving what looks like a nasty burn, but treatment with antibiotics is usually effective. However, another of *S. aureus*'s toxins, enterotoxin B, has the potential to cause much more harm. If enterotoxin B is recognised by an individual's immune system it can trigger an extreme reaction. An overreaction by the immune system can cause 'toxic shock syndrome' – a sunburn like rash appears, blood pressure falls and multiple organs can fail, leading to death.

Skaar proposed to treat this thankfully very rare condition with a light-sensitive molecule that is absorbed by the bacteria. When a specific wavelength of light is shone on the affected skin, the molecule is activated and the bacteria destroyed. Though this specific treatment is still in the experimental stages, there are other forms of light therapy that are in use and can be effective in treating conditions as varied as skin cancer and macular degeneration.[‖‖]

Drugs based on the porphyrin structures we met earlier, that cause all the problems in porphyria, can be potentially put to good use as medicine. Porphyrins have been used, because they are known to absorb light very effectively. The chemical reactions they undergo as a result of absorbing this energy can lead to cell damage; the trick is to direct this cell damage to the right place. By chemically modifying porphyrins, they can be specially tailored to absorb specific wavelengths of light. These light-sensitive molecules can be introduced into the body by swallowing or injecting them. They then get taken up by rapidly proliferating tissue and are therefore concentrated in cancer cells, or the blood vessels that grow behind the retina causing loss of sight in wet macular degeneration. These cells

[‖‖] Age-related macular degeneration is a condition where the central region of vision is slowly lost. It usually begins to affect people in their fifties and sixties.

can then be destroyed by shining a specific wavelength of light on them.

Porphyrins have been known to dramatically increase the skin's sensitivity to sunlight since the early twentieth century. In 1913, the German scientist Meyer-Betz saw his face swell up to an alarming size after he injected himself with haematoporphyrin and then sat at the window seat on a tram. He was still sensitive to sunlight two months later. Photodynamic therapy had an even earlier start, in 1903, when Niels Ryberg Finsen was awarded the Nobel Prize in Medicine or Physiology for his treatment of the skin condition lupus vulgaris with the light from a carbon arc lamp.*** Porphyrins and light therapy were combined in the 1970s in human trials to treat skin cancer. Since then, photodynamic therapy has been successfully applied to treat a wider range of cancers and other conditions, and has become a more mainstream part of medicine.†††

Perhaps this could explain a vampire's sensitivity to sunlight. Vampires, with their extreme blood-drinking habits, are likely to have excess haem, or related porphyrins, in their blood, just as those undergoing photodynamic therapy do. And, rather than single controlled doses that target a specific wavelength of light, they would be ingesting large amounts of a variety of porphyrin-based compounds on a nightly basis. A variation of this theory was used by the science fiction writer Richard Matheson in his classic vampire novel.

In the story that introduced this chapter, from *I Am Legend*, Matheson wrote about light being used to treat a bacterial disease. The novel was written in 1954, showing he was ahead of his time. The disease in Matheson's novel happened to cause a lot of the symptoms we associate with vampires. Following simple logic, if these vampiric symptoms were the by-product of a bacterial infection and bacteria are killed by UV light, then vampires can also be killed by sunlight. The sunshine treatment was effective, in that it was fatal for the bacteria, but, unfortunately, it was also fatal for the patient.

***Lupus vulgaris is an infection of the skin caused by the tuberculosis bacteria.

†††After much research and refining, a range of compounds are now used in photodynamic therapy, not just porphyrins.

On the face of it, Matheson based his science fiction on some solid science fact. That bacteria are killed by UV light is certainly true; specifically, UVC light. UVC is of sufficiently high energy to be absorbed by nucleic acids – the basic units of DNA – and rip them apart. This narrow band of light can be used to effectively chew up the DNA in bacteria, and without properly functioning DNA, the bacteria dies. As we saw earlier in the chapter, UVC from the sun does not make it to ground level here on Earth because it is absorbed by our atmosphere, meaning Matheson's vampires and their bacteria would be safe out in daylight.

In real life, UVC lamps are used to sterilise food and hospital equipment, but using those same lamps on something the size of a human would not work. Humans are dense, and the UVC light simply won't penetrate deep enough to reach all the bacteria in the body. There is also the problem that any light that reaches our own cells will also rip apart our DNA just as effectively as bacterial DNA, explaining why sunlight also kills the host in Matheson's story. But this is picking holes in what is generally a well thought out vampire theory, which is used extremely effectively in a tense novel.

Matheson didn't invent the light aversion characteristic in vampires, it was already established by the time he came to write *I Am Legend*. It didn't come from folklore vampires or the nineteenth century fictional vampires such as Lord Ruthven and Dracula either. Somewhere between 1897 and 1954, vampires went from a predominantly nocturnal lifestyle to running from the sun in fear of their lives.[‡‡‡] Death by sunlight has become one of the most well-known and most repeated characteristics of vampirism, but it had to start somewhere. For those who don't want to know the truth, look away now.

5.8 YOU CAN'T HANDLE THE TRUTH!

The start of a vampire's extreme reaction to sunlight can be dated very precisely, to 1922. Before this date, vampires were simply nocturnal (mostly) because that is when we expect creepy and unnatural creatures to be out and about doing creepy and unnatural things.[§§§] After 1922, vampires could be destroyed

[‡‡‡]Or should that be deathstyle and fearing for their unlife?
[§§§]See also witches, werewolves, ghosts and all things that go bump in the night.

by sunlight and this single well-known characteristic can be attributed to one film, called *Nosferatu*, directed by F. W. Murnau.

Nosferatu was the first full-length vampire film ever made. Murnau had wanted to film *Dracula,* but he didn't want to pay for the rights, so he changed the names of the characters, the locations (away from Transylvania and Whitby to Bavaria) and, most importantly, the ending. Despite the changes, this film was perhaps one of the cinematic depictions that is more faithful to the *Dracula* novel than the majority of those that came afterwards.

The biggest deviation from the original story comes at the very end of the film. Having sated himself on the blood of Ellen Hutter (Murnau's renamed Mina Harker), Count Orlok (Murnau's renamed Count Dracula) hears the cock crow. Perhaps thinking about how he has missed so many dawns, he moves over to her bedroom window to watch the sunrise. As the sun comes up over the houses it falls on Count Orlok. In a moment, of what was for the time a spectacular camera effect but today looks rather cheesy, Orlok disappears in a puff of smoke.

Vampires being destroyed by sunlight is the invention of one film maker who wanted to show off with an impressive special effect. An almost forgotten film has had an extraordinary impact on popular culture. Special effects may have improved over the years but vampires are still, almost always, destroyed by sunlight.

CHAPTER 6

Supernatural

> 'I saw the whole man slowly emerge from the window and begin to crawl down the castle wall over that dreadful abyss, face down'.
>
> Bram Stoker, *Dracula*

Louis was twenty-five years old when he became a vampire. As his human life ended, his vampire life began. The vampire blood he had drunk renewed and improved on his previous human existence: he said, 'it was as if I had only just been able to see colours and shapes for the first time.' His senses were heightened far beyond what a human was capable of experiencing. He found he was much stronger than before and he could move so fast, humans saw him pass by as only a blur.

Louis, Lestat and Armand, are the central vampires from *Interview With the Vampire*, the first novel in the series of 'Vampire Chronicles' written by Anne Rice between 1976 and 2018. These vampires took much of their inspiration from the glamorous tall, pale and handsome revenants of nineteenth-century fiction. Rice then enhanced these seductive creatures with supernatural powers. Count Dracula may have been very strong, and had impressive powers of mind-control, but this new breed of vampires were able to run faster than a speeding bullet and leap tall buildings with a single bound. However, they were most definitely not here to save the day. These anti-superheroes

Vampirology: The Science of Horror's Most Famous Fiend
By Kathryn Harkup

Published by the Royal Society of Chemistry, www.rsc.org

(or is that super-anti-heroes?) were only interested in themselves and their need for human blood.

Vampires have been attributed with impressive powers for centuries, but these supernatural abilities appear to have grown and multiplied since the beginning of the nineteenth century. Their enormous strength, speed, mind control, and the ability of some of them to climb up walls or even fly, would be the envy of any self-respecting superhero. However, most superheroes or villains only get one, or maybe two superpowers. Dracula and his fellow blood-suckers appear to have several at their disposal. Collectively, their diverse supernatural talents outshine any and every superhero and supervillain from all the comics, graphic novels and films, but vampires have rarely been included in these stories. When vampires have been included, it is usually safe to assume they are the bad guys. Though there have been a few notable exceptions to the 'any vampire is a bad vampire' rule, for example, Blade and Angel, who were both on the side of humans, but they were also both atypical vampires.[†]

Dracula was a more typical example when it came to vampire superpowers. In the novel, as Van Helsing and his fellow vampire hunters got to know more about the Count, the list of his extraordinary abilities grew. He could control the weather and hold people under a hypnotic spell using some kind of extra sensory perception. It made him a formidable foe but, as with many superheroes or villains, his superpowers were also his undoing.

This chapter examines whether there could be any scientific basis for some of the supernatural abilities demonstrated by vampires. Could there be any tricks or adaptations that vampires might have acquired to out-compete humans in so many ways? Or, perhaps, there are other reasons vampires have been credited with these superhuman attributes.

Though there are many characteristics, both physical and mental, shown by these revenants, that humans simply cannot come close to matching, there are always people with special talents and abilities. Some humans have sought ways to enhance their physical performance, and some of the abilities and traits that are sought after in elite sports-people might also be desirable for a vampire.

[†]Blade was part human and Angel had retained his soul.

6.1 STRENGTH AND SPEED

Some humans are clearly faster and stronger than others. Top athletes can run faster and further in shorter amounts of time than the vast majority of the population. To achieve these feats, they have to train hard and eat huge amounts of calories to supply them with the energy they need. A vampire lifestyle would seem to require the rapid and sudden physical exertion of, for example, sprinters or boxers, rather than endurance sports such as marathon running. They would also need to be very alert and able to concentrate during their nightly business of drinking blood to avoid being caught. Alertness and concentration are skills perhaps associated with sports such as top-level tennis or Grand Prix racing.

Lying around in a coffin all day is hardly a rigorous training programme for building up muscle strength, but by contrast it might be a good time for meditation and training to focus the mind. Nightly excursions to feed, over the period of years, and perhaps centuries that are apparently available to modern vampires, would certainly give them time to hone their technique. A vampire's night-time tracking and attacking of their victims is perhaps sufficient physical exercise to keep them trim. Those examples of the undead who did not acquire the necessary skills sufficiently early on in their vampiric career would be quickly found out and eliminated. Top athletes also exist in a highly competitive world but, while losing the top spot in a championship or world rankings can be devastating, it is not as lethal as the consequences for a vampire who fails to perform at their best. Perhaps this accelerated survival of the fittest would explain the spurts of rapid vampire evolution outlined in Chapter 1.

In addition to hours of training and practice, top athletes carefully prepare their body to be able to get the best from it. Diets are meticulously controlled to give a balance of nutrients and sufficient calories to produce optimum health. Vampires, on the other hand, seem to exist on a very meagre diet in terms of both calorie intake and balanced nutrition (see Chapter 3), so they are likely to need some supplements to help their body achieve the great physical exertions they are apparently capable of. Perhaps, like athletes, vampires would have to take supplements and additives to optimise their nutritional intake. There are, however, no superfoods or miraculous supplements that can boost the body's performance to the extreme levels demonstrated in vampire films and books.

After months and years of training, an athlete's body must make specific and dramatic changes when it comes to the time of action. At the start of a race or the beginning of a match, adrenaline will start to flow through the body in anticipation of what is about to happen. Adrenaline is a hormone that triggers the 'fight or flight response' in the body. This hormone primes the body for action by diverting resources to where they are needed. The pupil of the eye dilates to allow as much visual information as possible to be acquired. Digestion is slowed so that resources can be diverted elsewhere. Heart rate and breathing are increased and, as a consequence, more oxygen is drawn in to the body and the heart works faster to deliver it to the muscles. Adrenaline also triggers the body's stores of glycogen to be converted into glucose, so that the oxygen and glucose can be combined to increase respiration and release more energy.

Even those who haven't trained can access the potential possibilities of adrenaline. Humans have, on occasion, shown extraordinary physical abilities or endurance without years of training. Every now and then a story emerges of a parent who was able to lift a car to free their trapped child from underneath the vehicle.[‡] These stories are anecdotal and testing them scientifically under controlled conditions is impossible, from an ethical as well as practical point of view. Nevertheless, some scientists, using their knowledge of human physiology and muscle mechanics, have speculated on whether these feats might be possible or if the reports are an exaggeration of what really happened.

The theory goes that 'hysterical strength' might be possible because we rarely use our muscles to their full capacity. Evidence from electric shocks demonstrates how a person can be thrown across a room, not by the electricity itself but by the contraction of the muscles triggered by the electric shock. In extremely stressful situations, such as finding your child trapped under a car, the adrenal glands release a flood of adrenaline allowing muscles to operate at their maximum capacity, and seemingly superhuman levels of strength can be exerted. But this is a very

[‡]The sight of a woman lifting a car off her trapped child was said to have inspired the writer Jack Kirby to create the Hulk.

short-term effect used for emergency situations. Though the effects of adrenaline and the potential strength of human muscles are well understood, it is still not clear if 'hysterical strength' is a real phenomenon. Perhaps, if it is real, vampires could have easier access to this heightened physical state. However, there are considerable downsides to this theory.

High levels of adrenaline in the body over long periods of time are not good for us; just look at the detrimental health effects of long-term stress – from trouble sleeping to heart problems. In extreme cases, it can kill. The phenomenon of excited delirium syndrome has been recognised in recent years to describe incidents where people have certain types of agitated and aggressive behaviour before their sudden death. Law enforcement is often called to deal with the behaviour of these individuals but they are often unresponsive to verbal instructions and resist arrest. They exhibit 'superhuman strength' and are impervious to pain when officers try to physically restrain them. These symptoms are all characteristic of a lot of adrenaline flowing through the body. Officers also commonly note how unusually hot the person is, and they have often been called to the scene because the individual is wandering around wearing very little considering the local temperature.

After this period of extreme agitation, the distressed person can often become calm very suddenly, followed by rapid unconsciousness and death. Autopsies do not show any physical cause for the rapid decline in health or explanations for why the individual died. Research is very limited in this area but there have been links made to cocaine use as well as severe mental health issues. One theory is that the syndrome is related to abnormal release of dopamine in the brain. This is a compelling theory because it would explain the coincidence with cocaine use and severe mental illness. The dopamine receptors in the hypothalamus are also responsible for regulating body temperature. This is an extreme and rare example of how the human body can exert surprising strength and how serious that can be for a person's health. It also puts Renfield's comments in *Dracula* in a different light when he asserted that he believed 'madmen' had unnatural strength.

A vampire would have to be able to tap into their body's muscular potential more easily, but would need a better way of dealing with the detrimental affects of high levels of adrenaline in their body. For athletes, being able to sustain high levels of exertion is where the training comes in, enabling them to operate their body at high levels of energy consumption without having to rely on a continuous supply of adrenaline.

Without an intense training programme, could there be other ways a vampire might develop greater strength and speed than humans? The example of the electric shock shows how an external stimulus could be useful for strength, but these muscle contractions aren't directed in any way. What is needed is controlled strength. Could a vampire simply have stronger muscles?

Other animals, closely related to humans, show increased strength for their size. Chimpanzees, though much smaller than humans, are typically stronger. There has long been anecdotal evidence of the enormous strength of humanity's closest living relative, but there have been few detailed studies. Occasional stories of captive chimpanzees lifting huge weights, often during moments of stress, are not a reliable way to compare the strength differences between the two species. It could be these chimps are showing the same 'hysterical strength' that is anecdotally shown by humans.

There are, of course, differences between chimpanzee and human body shapes, in both the skeleton and muscle arrangements, which make our close cousins more suited to short periods of more powerful activities, like tree climbing, and for us to endure long periods of walking or running. These anatomical differences explain a large part of the chimpanzees perceived strength. Vampires obviously resemble humans more than they do chimpanzees, so changes in their physique can't explain their increased strength, but perhaps there are other insights to be gained from chimps.

In 2017, a study looked at the muscle tissue from chimpanzees and compared it with humans and other animals. They found that there were significant differences in the type of muscle fibres that made up the muscles in different animals. Human muscle was dominated by slow twitch fibres that allow more sustained or endurance activities. By contrast, chimpanzee muscle was found to have a mixture of slow and fast twitch fibres, giving them a greater capacity for short-term activity and also increasing their strength. They estimated that the different fibre composition of chimpanzee muscles, together with the fact that these muscle fibres were longer than in humans, gave them as much as 1.35 times the strength of the average human.

Perhaps part of a human's transformation into a vampire could involve changes to muscle fibre. An increase in fast-twitch muscle fibres may explain some of the extra strength demonstrated by vampires in fiction and films, but it would not be enough to allow them the strength and speed demonstrated by vampires like Lestat and Louis in *Interview With the Vampire*. Are there other ways to get the most out of the body?

6.2 GAMING THE SYSTEM

Striving to be the best and gain any advantage over competitors has driven some athletes to try all sorts of things to improve their performance. By studying how the body carries out the process of respiration naturally, we can look for ways an athlete or a vampire might be able to improve on these basic human systems to achieve a boost in energy production that would be needed for the extreme speed and strength. There are many components necessary for the body to carry out respiration, and therefore several ways to increase energy production without the potentially damaging effects of adrenaline or other drugs.

To increase the amount of energy that is available, first a larger amount of oxygen needs to be circulating in the bloodstream. Hypoxia-inducible factor (HIF) activates erythropoietin (EPO), which, in turn, stimulates red blood cell production. With more red blood cells available, more oxygen can be transported round the body. Athletes have been known to take extra HIF to improve their performance and so checks are carried out. The level of HIF in the blood can be tested close to race day as well as during training periods for comparison for what a 'normal' level would be for that individual. It is therefore possible to check that an athlete has not been trying to increase their red blood cell levels artificially. Perhaps vampires naturally produce more HIF or EPO to sustain their increased oxygen needs.

Alternatively, perhaps vampires are not relying on increasing their own red blood cell count and are instead making use of the blood they ingest every night. A vampire is not going to be short of red blood cells, thanks to their dietary habits, but these cells are in the wrong place. Part of a vampire's adaptation could be to improve absorption of whole red blood cells into the circulatory system for its own use before they are broken down by digestion.[§] Some athletes have attempted something similar to improve their body's normal performance in competition. Athletes, rather than drinking other people's blood, have been known to drain some of their own blood and store it. They then let their body build up their normal levels of red blood cells, but on race day they transfuse back in the extra red blood cells they have been saving up. This is known as autologous blood doping and is strictly against the doping rules in most sports.[¶] However, it can be difficult to prove unless fridges full of chilled blood are discovered.

Even with more blood in their system, a vampire also needs to get more oxygen into their lungs to carry out increased respiration. Oxygen can be absorbed through the skin, but it is not nearly enough and it will only penetrate a tiny fraction of the way into the body before it is used up. Anything deeper inside the body would be starved of oxygen and die. And although vampires are supposed to be dead, they must breathe at least once in a while to enable them to have air to pass over their vocal chords to talk. Lung capacity can be increased with training. For example, divers, singers and those who play wind instruments, can draw in a greater volume of air with each breath. Perhaps vampires have a similar increased lung capacity for when they need more oxygen. Alternatively, they could make greater use of a different method of extracting energy from glucose; one that doesn't require oxygen.

Humans doing a lot of exercise can reach a point where the air they are drawing into their lungs is not enough to meet the energy demands of their muscles. Muscle cells deprived of oxygen have a backup system of getting energy from glucose without oxygen. Anaerobic respiration is a temporary measure that releases energy, but also leads to the build-up of lactic acid and causes the cramps felt when muscles are exhausted from physical exertion. Drugs that alleviate pain can allow someone to push their body much further, perhaps vampires are naturally impervious to the pain of lactic acid build-up or, like some other creatures, have a natural ability to operate anaerobically for a long period of time.

[§]I don't know how this might be achieved as any 'gaps' big enough to allow red blood cells to move from the digestive system into the circulatory system would also let in a lot of other unwanted things. It would also allow blood out.

[¶]This was apparently quite common in competitive cycling in the 1980s and 1990s and was known as 'vampire tactics'.

Nereid worms, and a few other species, use the lactic acid pathway to derive much of their energy and these abundant creatures, often used as bait by fishermen, seem to thrive on it. But there are some creatures that take it even further and live entirely without oxygen, such as *Clostridium botulinum* (the bacteria that cause botulism in spoiled food) and three species of loricifera (tiny marine creatures that live in sediment). The vast majority of these anaerobic creatures are single-celled organisms. The three species of locifera are a rare exception because they are multicellular. Loricifera process hydrogen using specialised hydrogenosomes (evolved from mitochondria) to obtain their energy, instead of oxygen or lactic acid pathways. But for these few, very tiny exceptions, multicellular, large-scale complex life is simply not feasible on an anaerobic system – as far as we know – because these alternative systems simply do not release the same amount of energy as aerobic respiration. For example, the lactic acid process releases about 5% of the equivalent reaction if oxygen was present.

However, some larger animals, such as freshwater terrapins, can survive an impressively long time without oxygen. In addition to being able to carry out extreme physical activity, better oxygen and energy management would be an important consideration for a vampire that spends much of the day underground in coffins with a limited air supply. Coffins contain approximately enough air to support human life for around twenty minutes. Though it might not seem like it, being able to survive many hours in a low oxygen environment is a supernatural ability for a human, but not some animals. Could vampires be more like terrapins or naked mole rats than humans?

Terrapins can survive for two days at room temperature without oxygen, but for up to six weeks under the ice in frozen ponds. These and other animals that go for long periods without oxygen, or spend much of their life in environments with low oxygen levels, such as naked mole rats living in poorly ventilated underground tunnels, have had to adapt considerably to compensate for the damage caused by lack of oxygen to the body.

Cells such as neurons (nerve cells) guzzle energy at a fantastic rate to carry out their normal functions. Cut off the power supply and damage occurs rapidly because molecular pumps, which keep everything in the right place, stop working. The release of chemicals from the cell is uncontrolled and can spread to adjacent cells, causing further damage. When oxygen is later restored, these cells produce harmful by-products, such as reactive-oxygen species, that go on to do yet more damage. In humans, the result is a stroke.

Other animals prevent damage by having a modified metabolism, so fewer reactive-oxygen species are produced and specialised proteins are made that protect the cells. By contrast, naked mole rats don't modify their metabolism and tolerate a high level of cell damage. They live long, healthy lives because the processes that stop damaged cells from dividing are amplified. What is all the more interesting is that humans have the same genetic instructions in their body that would allow us to be more like terrapins or naked mole rats, we just use them in a different way. If humans, or vampires, could tap into these genetic instructions and switch certain genes on at the right time and the right way, we could have more resilience to shortages of oxygen.

We all have the genetic potential to be superhuman but how genes are controlled, and what this might mean for humans and vampires, will be explored more in future chapters. But, for the time being, there is more to investigate regarding the ability of some animals to perform feats that are impossible for a human, but are apparently par for the course as far as vampires are concerned.

6.3 CLIMBING THE WALLS

One of the most dramatic moments in *Dracula*, the moment when all of Jonathan Harker's fears about the strangeness of his host at Castle Dracula are confirmed, is when he sees the Count climb out of a high window and crawl down the outside vertical wall of the castle. It is not something humans should be able to do without specialist climbing gear, but Count Dracula has no trouble navigating the sheer descent with nothing more than his hands and feet (see Figure 6.1).

Some animals appear to defy gravity easily. Snails and slugs can use slime to form a sticky layer that enables them to climb vertical walls. Octopuses use suction cups in their tentacles to

Figure 6.1 Cover of *Dracula* from 1901. Archibald and Constable Company.

adhere to surfaces. Several species of insects and lizards can crawl up walls with apparent ease and without the need for slime, suction or specialised equipment. The only human that seems capable of these feats, however, is the fictional Spider-man and he got his power from a radioactive spider. Stoker described the Count crawling down the wall like a lizard. Could Dracula have borrowed a trick or two from geckos?

In order to climb walls and walk across ceilings, geckos and spiders rely on tiny specialised hairs on their feet, known as setae. These hairs branch into smaller and smaller filaments until they are hundreds of times thinner than a human hair.[||] Thousands and thousands of tiny bristles, packed onto the feet of these creatures, mean they have a huge surface area for the size of their foot pad. With such a high surface area, a certain type of force comes into play. This is nothing to do with 'The Force' found in a galaxy far far away, but a tiny attraction between atoms and molecules called Van der Waals forces.

The atoms that make up molecules have a positive nucleus surrounded by a cloud of negative electrons. Some atoms attract this negative cloud of electrons more easily than others, meaning that across a molecule made of several atoms there will be patches that are more positive than others and some that are more negative. In addition to these permanent positive and negative areas, there can be temporary effects. As the cloud of negative electrons moves and vibrates around the positive core of an atom, there can be occasional moments where electrons are more concentrated in one area than another. This means that one area on the atom can, for a fleeting instant, be more negative than another area. These transitory positive and negative distortions on one atom can interact with distortions on another atom. Opposites attract and like charges repel each other. Van der Waals forces are the sum of all these tiny attractions and repulsions. Over billions of trillions of atoms, this very weak force can build up to give impressive results.

With a gecko's foot pads, the weak Van der Waals attraction is between molecules at the tips of the tiny setae fibres and molecules in the wall or ceiling that a gecko is climbing on. Every square millimetre of each gecko toe has around 14 000 setae. If the animal keeps its feet clean, and gecko feet are self-cleaning, it can stick to walls indefinitely as the stickiness never wears off or wears out. A gecko uses less than 5% of its total surface area to support its whole body weight. In fact, the setae are so good at sticking to smooth surfaces that a gecko can hang its whole body from just one finger or toe. Using all its setae, an average 70 g gecko could support 133 kg in weight.** This may seem like overcompensating,

---

[||]In *Dracula*, Jonathan Harker does comment that the Count has hairy palms, though setae are too fine to be seen with the naked eye.

**Please do *not* try this with your pet gecko.

but geckos do not live in environments where all surfaces are smooth as glass, they must be able to cling on to very rough surfaces as well as very smooth – much like Dracula clinging to a stone wall, rather than the smooth glass of a window.[††]

Spiders use the same Van der Waals trick as geckos, but cover less than 1% of their surface area with setae to achieve the same wall-climbing and ceiling-clinging abilities. They can get away with much less coverage because they are generally much lighter than geckos and have eight feet to spread the load, rather than four. Scaling up these effects to human size presents problems.

As mass increases, surface area decreases proportionately. A spider needs 0.9% of its surface area to be covered in setae. Scaling that up to the size and mass of a gecko, 4–4.5% coverage is required. Scaling up further, to the size of a human, 40% coverage of the surface area is needed, and that surface has to be in contact with the wall. But, as we have seen, there is a certain degree of overcompensation in the gecko.

In 2014, DARPA (Defense Advanced Research Projects Agency) showed that a 218 lb researcher carrying a 50 lb load (117 kg total) could climb a vertical glass wall using only two small paddles covered with synthetic gecko material. The climbing action might not have been as elegant as a gecko or spider, but it showed that the weight of an adult human could be supported by just one gecko paddle as the researcher climbed. However, few details of what the paddle was made from or how big it was were released.

The following year, another team based at Stanford University made a separate attempt, but released a few more details about the process. Their two fake gecko footpads were each 140 cm^2 and made from polydimethylsiloxane formed into millions of neatly arranged microwedges, each approximately 100 μm tall (twenty times the diameter of a single gecko setae). The paddles could attach to a vertical glass wall simply by applying the climber's weight to the pad. These pads could be released by removing the applied pressure, allowing a 70 kg researcher to take hundreds of steps up the glass without a single failure. It illustrated the overcompensation that geckos need to manage the different surfaces they would encounter in the wild. The trick was having a

[††]The only surface a gecko really struggles with is Teflon because it has a very low surface energy.

uniformity to their synthetic gecko material that gave a lot of contact with the wall and a smooth surface to climb on.

Dracula did not have a smooth glass wall to climb down, but a rough stone castle wall. The gaps in the wall between the stones may have offered some extra opportunities for the Count to get a finger-hold on his way down the wall. His long sharp nails may have helped or, perhaps, instead of using his own hands to grip the stone, he had specialised gloves. The use of high-tech gloves would not necessarily be out of keeping in the novel, though it is hard to figure out where a centuries-old count living in a remote part of Europe would have got hold of such equipment.

Stoker had his titular vampire steeped in tradition and superstition, but his vampire hunters were using up-to-date science and technology – from blood transfusions to state-of-the-art (for the time period) voice recording equipment. Van Helsing and his team of vampire hunters used both the latest gadgetry and traditional methods to overcome the Count and his world of superstition and old beliefs. In some ways, the novel can be read as how the Victorians simultaneously made huge scientific and technological advances, while also maintaining a fascination for the occult and supernatural forces. This overlap between what is now accepted as science, and pseudo-science, was most apparent in Victorian investigations into the mind and its abilities. Even today, the human mind holds many mysteries and the boundaries between some aspects of brain science and the supernatural can appear very blurred.

6.4 SCIENTIFIC AND PSYCHICAL RESEARCH

Stoker was evidently interested not only in technology and new scientific discoveries, but also the Victorian craze for more unorthodox theories such as mesmerism and spiritualism. In the nineteenth century, investigations into what are now considered supernatural forces were considered much closer to conventional scientific research. The rapid advancement of scientific understanding in the previous century, as part of the Enlightenment, had revealed incredible new phenomena such as electricity and a whole new array of previously unknown chemical elements. Invisible forces that could influence human health and well-being did not seem so far fetched. Spiritualism and investigations into the occult were all the rage in Victorian England and nineteenth-century writers made the most of it. The vampire fiction from this era is full of examples of mind control and psychic connections, but none more so than *Dracula*.

In the late eighteenth century, Franz Anton Mesmer put forward the idea that there was an invisible force contained within every living thing. He asserted that this animal magnetism, as he called it, could be manipulated by magnets to bring about medical cures. When Mesmer's fantastic claims came under closer scientific scrutiny, however, there was a distinct lack of evidence for this mysterious force. An investigation carried out by members of the French Academy of Sciences, which included Benjamin Franklin, the well-renowned scientific scholar and statesman, dismissed the idea of a healing force. Nevertheless, the dramatic effects that could be produced on individuals by mesmerists or magentisers, as they became known, were difficult to ignore. In the 1840s, the Scottish surgeon James Braid, convinced that a powerful influence could be asserted by a mesmeric operator on an individual, introduced the word 'hypnotism' and promoted it as a useful medical technique and potential cure for all manner of conditions.

The term 'animal magnetism' quickly fell out of favour amongst scientists, but many aspects of this research into the unknown were not so easy to dismiss. Many of the reported phenomena, such as action at a distance; for example, reading the pages of a closed book, and manipulation of human behaviour, appeared to overlap with physics and psychology, the more orthodox branches of science. Invisible forces that acted at a distance may seem fantastical, but well established scientific phenomenon such as magnetic fields can also cause metal objects to move without any visible connection between the object and the magnet. The nineteenth-century interest in mental health, and especially nervous disorders, chimed with a hypnotist's influence on the behaviour and temperament of his subject. The boundaries between science and pseudo-science were somewhat blurred. Spiritualism was considered outside the bounds of serious science, but calling it psychical research brought it nearer the fringes of acceptability. Mesmerism and animal magnetism were definitely unorthodox, but hypnotism was on the border of orthodox scientific research and practised by some well-renowned physicians.

Many practitioners of psychical research advocated keeping an 'open mind'. Bram Stoker personified the close but uneasy relationship between science and superstition in the character of Van Helsing, 'a philosopher and a metaphysician, and one of the most advanced scientists of his day; and he has, I believe, an absolutely open mind.' And, in that same spirit, could there be anything to the extraordinary mental powers of the undead?

6.5 MIND CONTROL

It is generally acknowledged that vampires have incredible powers to control the minds of their victims. There are also examples of blood-sucking revenants who use their powers to control the weather, at least in their immediate environment. Others, like Dracula, can also control the actions of other animals, such as rats and wolves, like an undead Doctor Dolittle. The most common form of a vampire's superior mental abilities, however, is used to hypnotise their human victims and overpower their revulsion to being bitten. Let's examine each of these incredible abilities one at a time.

The weather-related form of these extraordinary powers is the most difficult to explain scientifically. This impressive talent could be interpreted as a form of telekinesis, or psychokinesis, which is the ability to control and move material objects by the power of thought. In the example of Count Dracula, the objects that are manipulated are tiny droplets of water in the form of fog, or molecules of air to create breezes and winds to waft boats more speedily to their destination. The objects being moved may be small, but there are an awful lot of them and each will require a force to be applied to them. There is no known scientific mechanism that could explain how such a force could be created, let alone controlled, by mental processes. Fantastic sums of money have been offered to anyone who can demonstrate this ability, on even a modest scale, in a controlled experiment. To date, no one has made a successful claim.[‡‡] So much for Dracula

[‡‡]There is a wikipedia page, https://en.wikipedia.org/wiki/List_of_prizes_for_evidence_of_the_paranormal, listing all the prizes currently available for anyone who wants to have a go.

being able to create his own private storms. What about the other claims made about vampiric mental powers?

Vampires in literature seem to have an incredible power to control their human victims. The level of control varies, from holding an individual in some kind of hypnotic trance, preventing them from defending themselves, to a more specific interaction. For example, in *Dracula*, the Count and Mina drank each other's blood, and from that point on they were linked in such a way that they each knew what the other was thinking and seeing, even though they were separated by hundreds of miles. Dracula and Mina had a private psychic link that no one else could access.

The first step is to use the mind to lull the victim into a false sense of security. It is well known that some people have powerful and persuasive personalities; from controlling partners to product placement and nudge advertising, people can be influenced to think in a certain way. Such control is built up over a period of time. Vampires seem to be able to exert a much greater level of control and establish it over a much shorter timespan, almost instantly, much as hypnotists can exert control over their subjects.

Hypnosis is a trance state, where a person is very relaxed but open to heightened imagination and becomes suggestible. It is often compared to daydreaming or losing yourself in a good book or film, where you are fully awake but scarcely aware of the real world around you and very focused on the book, film or person hypnotising you. The mind undoubtedly has the ability to tune out external stimuli whilst remaining alert and focussed, but highly suggestible. Quite how the brain achieves this is not known. We may understand a great deal about how the brain functions, but there is still no great theory of mind that would help explain how hypnosis works. One of the most widely accepted theories is that hypnosis is a way of accessing the subconscious mind, the part that controls the everyday mundane activities that we don't have to actively think about, such as breathing or walking. During hypnosis, it is thought that the conscious mind has a rest while a hypnotist gains control of the subconscious. It certainly explains why people under hypnosis tend to be uninhibited; the barriers put up by the conscious mind have been bypassed.

While there are many mysteries relating to how the mind works, science can offer some insight into how the body and brain behaves during hypnosis. Heart rate and breathing slow, probably because of the relaxed state of hypnosis. Electrical activity in the brain can be monitored by electroencephalographs (EEG) and shows that the brain produces different brainwaves during hypnosis, closer to those seen during sleep and dreaming. Looking at the cerebral cortex, the outer layer of the brain involved in conscious thought, shows increased activity in the right hemisphere, associated with imagination and creativity; while there is a decrease in activity on the left side, associated with logic, deduction and reasoning.

Hypnotism has been used for a variety of applications, from entertainment to treating addiction and phobias, or even to ease pain. It would certainly be of benefit to the average vampire, who must overcome their victim's fear of an undead creature suddenly appearing in the middle of the night, and to stop them crying out in pain when they are bitten. In the nineteenth century, vampire bats were believed to have something like hypnotic powers because they were thought to lull their victims to sleep with the beating of their wings. In actual fact, it was simply that the bat's target was already sleeping and the bite from a careful bat was not painful enough to wake the animal. Maybe vampires use similar methods to the animals they are so often associated with, either by the supposed hypnotism or simply waiting for their victim to fall into a deep sleep.

Hypnosis might explain a vampire's influence over a victim when they interact directly, and as a phenomenon demonstrated by many humans, it can hardly be classed as a superpower. But what hypnosis doesn't explain is the psychic link between Mina and Dracula that enables them to share detailed information, even though they are separated by several hundred miles. When the Count flees England for his native Transylvania, the group of vampire hunters are able to track his movements using Mina's special psychic link. Though many Victorian mesmerists claimed their influence extended over a considerable distance, and individuals had knowledge of things they could not see, such as the time on a watch held behind their back, these abilities were difficult to explain or prove in scientific terms. Such action at a distance seems to be more like extra sensory perception (ESP).

To exchange information privately between two people, there must be some kind of signal sent by one person that can be received and understood by the second person, and only the second person. To produce a signal that can reach hundreds of miles, the closest real-world equivalent might be broadcasting radio waves that are picked up by a receiver. Radio stations transmit a signal in every direction but not everyone detects that signal – a radio has to be tuned to a particular frequency in order to listen to that radio station. But, anyone who knows the frequency and has a radio can listen to the broadcast.

For a living thing to emit a signal that can reach hundreds of miles would require an enormous amount of energy. There would be no problem with detecting this energy; science has devised ways of detecting individual photons, and we can monitor even tiny fluctuations of electrical activity in human brains. If it is possible to identify a handful of neurons firing, it should be easy to detect an ESP signal being created by a vampire, even if the message cannot be understood. Yet, so far, no such signal has been detected. Even if vampires had developed a way to send these kind of signals and interpret them in the detail that is shown in *Dracula*, these would not be private signals. In practice, anyone should be able to receive the signals if they know the frequency.

Knowing what another person is thinking, even when they are close to you, is not always easy. There are tells to look for and conversation can guide someone's thoughts and elicit clues from replies. But this is not what Dracula and Mina achieved. Mina was fully able to access the Count's mind, seeing what he saw and hearing what he heard, and *vice versa*. Though it might seem incredible that one person could access the thoughts and experiences of another, science has shown ways that this might be possible.

To a certain extent, thoughts can be monitored using techniques such as EEG, mentioned earlier, and functional magnetic resonance imaging (fMRI). The first technique is a way of monitoring electrical activity in certain parts of the brain. It is non-invasive and works by placing electrodes on the scalp. The technique is most often used to diagnose abnormal patterns of activity that might be a result of a sleep disorder or epilepsy, it can also help determine the depth of coma or even brain death.

An alternative, non-invasive procedure is fMRI, which can measure blood flow or changes in metabolism in different parts of the brain. Using knowledge of which parts of the brain are responsible for different mental activities, it should be theoretically possible to work out what an individual is thinking about using these techniques. But there are problems.

EEG is relatively cheap and easy, and can detect electrical activity on the millisecond scale – almost real-time. The second technique, fMRI, relies on large, expensive equipment and it is slow; there are seconds or minutes of delay between the brain's activity and its observation. The information that can be gained from both these techniques is not nearly detailed or specific enough to identify individual thoughts, but that does not make them useless.

Some recent news stories have reported how paralysed people are able to control robots using only their thoughts and how people with locked-in syndrome can use their thoughts, detected by a machine, to communicate with the outside world. But it takes training to get these robots to work and even then it is in a relatively limited way. That is not to say there have not been some remarkable achievements in recent years. In 2019, using an implant placed above the region of the brain used to control movement, the electrical activity from this region could be read and interpreted by a computer. It allowed a quadriplegic man to control an on-screen avatar using only his thoughts. When he was placed in a specially built exoskeleton, he could use his mind to make the robot shell walk.

Brain implants are invasive and still require lots of training and practice to be translated into specific movements. An alternative system using non-invasive EEG allowed volunteers to control a robot arm that pointed and tracked a cursor moving randomly on a screen just by thinking. It is an impressive start, but the technology is still in its infancy. It is clunky, slow and mentally taxing, but it can be life changing for those who can use it to regain some independence. It is far from the detail and sophistication demonstrated by Dracula and Mina. It also requires complex machinery and powerful computing to gain these results. Technology will doubtless improve and bring increasing sophistication to the systems but this may be a long way in the future. Vampires have apparently been doing this for centuries and without the scientific knowledge and machines that enable current mind-control.

6.6 IF I COULD TALK TO THE ANIMALS

Sophisticated mental interactions between vampires and humans might be out of the question, but what about other animals? The animals that Dracula is supposed to have been able to control – bats, flies, rats and wolves – have varying levels of cognition, from simply acting on instinct, to the ability to learn commands and behaviours. These might be a simpler target for the mind control exhibited by the blood-sucking undead of popular fiction. Simple instructions like 'come here', 'move forward' or 'attack' are much closer to the level of sophistication shown by the mind-controlled robots described above. However, there is still no way of explaining how these messages are sent or received, and this places this ability firmly in the realm of the supernatural.

Vampires certainly have a very close relationship with particular animals, perhaps their ability to interact with them so effectively comes from their apparent ability to change into bats or dogs at will. The shape-shifting powers of many vampires, from the folkloric variety to modern vampires, is certainly a superpower but one that deserves a separate chapter. Read on to find out more.

CHAPTER 7

Shapeshifting

> 'I shall then give instances of persons thus afflicted, who were believed by others, and who believed themselves, to be transformed into beasts'.
>
> Sabine Baring-Gould, *The Book of Were-Wolves, Being an Account of a Terrible Superstition*

The vampires of the Caribbean are not like other vampires. They can disguise themselves by changing their shape into any form they choose. Usually, they travel around at night as a ball of fire. They sneak into houses under gaps at the bottom of doors or squeeze through keyholes to prey on the sleeping inhabitants. Victims are punched and pinched, but most often they have their blood sucked, leaving them weak and likely to die.

The vampire was introduced to the Caribbean in the seventeenth century by the French when they colonised Martinique. These vampires mixed with an African species brought over with slaves to produce a hybrid, the *soucouyant* (perhaps from the French 'sucer' – 'to suck'), and by the 1880s they were well established on the islands.

Quite apart from their distinctive appearance as floating fireballs, *soucouyants* have other attributes that distinguish them from their blood-sucking European cousins. *Soucouyants* are very much alive, but they are no less frightening than the undead. Their ranks are made up of people who have lived a bad life and

Vampirology: The Science of Horror's Most Famous Fiend
By Kathryn Harkup

Published by the Royal Society of Chemistry, www.rsc.org

who have made a pact with the Devil in return for being able to change their shape into almost any form they wish.

In order to acquire this shapeshifting ability, the person first has to go to a cemetery and dig up a fresh corpse. Then they extract the liver and use it to make an oil, which they rub all over their body to allow them to remove their skin any time between midnight and dawn. They hide their skin, usually in the hollow of trees or under mortars (large wooden vessels), and move around in the guise of a bat or other animal – but most often they choose to be a fireball.

Though *soucouyants* can change into any shape, they have to return to their human form during the day. If a human can find a *soucouyant* skin before the creature returns, sprinkling table salt (some say black pepper should also be added) on the underside of the skin will cause them to leap about in pain when they try to put it back on. A local priest can then be recruited to exorcise the demon spirit or, alternatively, the *soucouyant* can be roasted slowly over a fire.

Seeing a ball of fire floating around after dark is likely to leave a lasting impression on anyone. When almost any shape is apparently available to your average *soucouyant*, a floating fireball seems a little like showing off, and more likely to draw attention than divert it. Most other vampire-like creatures seem to prefer to keep a low profile so they can go about their human predations with as little fuss as possible. In that respect, being able to change shape is a huge advantage when it comes to disguise, and it can't be surprising that the ability is found in many of the vampire varieties described throughout history. As shapeshifting is such a widespread ability amongst the blood-sucking undead, is there perhaps more to it than whimsical story-telling?

First of all, the *soucouyant*'s method of shapeshifting is a recipe or process, rather than an inherent natural ability. Is there some lubricant that can help you transform your shape or slip out of your skin? The short answer is 'no'. Please do not go around digging up dead bodies and rummaging around in their abdomens for the liver. There is no substance within this organ[†] that can be incorporated or transformed into an oil that will allow you to shed your skin and change your shape, with or

[†]Or anywhere else, for that matter.

without the help of the Devil. Please don't do it. It is unpleasant, ineffective and illegal.

Most vampires do not even have to go through such dubious processes in order to change their shape or size, it is simply one of their many natural abilities. Ancient vampires, such as the Lamia (see Chapter 1) could disguise themselves as beautiful women to lure men to their bloody doom. According to folklore, vampires across various parts of Europe can transform into wolves, and a few other animals, though very rarely bats. Modern vampires seem to have expanded their shapeshifting repertoire to include bats and large dogs, or to simply deform their body shape to squeeze through the crack of a crypt door, or disperse themselves into a fine mist to escape the clutches of vampire hunters. In *Carmilla*, the titular vampire is seen as a large prowling animal. In *Dracula*, Van Helsing reveals that the Count can grow bigger or smaller at will, although he isn't observed doing this. Mina Harker, in her vampire form, is observed flattening herself to slide through the crack in the door of her tomb to reach her coffin.

This chapter looks at where these ideas might have come from and how a vampire might be able to achieve them. But, before we look at changing shape, what about changing size?

7.1 SIZE MATTERS

Changing your size is not as easy as swallowing the contents of a bottle labelled 'drink me', as in *Alice in Wonderland*, and it is a very different ability to that of changing shape. Some animals try to make themselves look bigger to appear more frightening, for example, cats arching their back and raising their fur, or puffer fish inflating themselves when startled. But it is an illusion: the cat relaxes and the fish spits out water to return to their normal size – the animal itself hasn't suddenly gained or lost weight. This does not seem to be what happens with vampires. Dracula didn't just take in a big lungful of air to swell himself up, or simply crouch to make himself appear small. His change in size must have been more dramatic than that, or Van Helsing wouldn't have commented on it. Such a change is difficult to achieve without rapidly gaining or losing mass and this would appear to break one of the fundamental laws of physics.

The amount of matter and energy in the universe is finite. In special circumstances it is possible to convert matter into energy and *vice versa*, but energy and matter cannot be conjured out of nothing or made to disappear like a magician's rabbit. Energy and matter, like the rabbit, are always somewhere, it's just not always obvious exactly where.

Mass and energy are linked by Einstein's most famous equation: $E = mc^2$. This equation shows that mass and energy are interchangeable, and states that energy (E) is equal to mass (m) multiplied by the speed of light squared. In simple terms, a tiny amount of mass is the equivalent of a staggeringly huge amount of energy. Or, alternatively, it takes a huge amount of energy to produce a tiny amount of mass. Dracula's rapid changes from big to small, or from a human sized creature to a bat, would release the sort of energy seen in atomic explosions. Growing bigger, or turning back from a bat to a full-sized vampire, would mean a gain in mass that would require similarly staggering amounts of energy. These rapid changes occurring in a London suburb would result in the total destruction of the city. This evidently didn't happen, as people would have noticed. Therefore, the transformation must have happened some other way that didn't wipe out the capital.

There may be a very simple explanation for the belief that vampires could grow bigger or smaller. It probably stems from the appearance of the dead bodies that were dug up during vampire epidemics in the eighteenth century. These bodies would have been in various stages of decomposition. As discussed in more detail in Chapter 3, the gases produced in the bloat stage of decomposition can make a corpse swell to up to three times its normal size; and dessication of a body buried in dry conditions can make it appear to shrink.

Vampiric transformations of size may have been taken from folkloric accounts of swollen or shrunken bodies that were then extrapolated into a superpower under the vampire's conscious control. But these post-mortem effects don't explain how vampires are thought to be able to transform into very differently shaped and sized animals. One way of avoiding breaking the laws of physics would be for these transformations to be some kind of disguise, mimicry or manipulation of the vampire's own body to give the illusion of a change in size or the appearance of another creature. Such extreme changes of appearance are unusual for humans and are more often associated with other species. Maybe vampires use some of the tricks evolved by some animals to avoid detection or to hunt their prey more successfully.

7.2 HIDING IN PLAIN SIGHT

Some animals mimic others as a means of defence or disguise. Several species of butterfly have patterns on their wings that look like owl eyes, to warn off predators. There are frogs and snakes that use bright colours to give the impression they are more dangerous and venomous than they really are. Animals such as stick insects, or grasshoppers that have evolved body shapes that resemble leaves, look like something innocuous so they are ignored, or are difficult to pick out from their surroundings. This method of camouflage – looking like something you are not – is called *mimesis*.

The better-known method of camouflage is trying to blend in to the background, and this is known as *crypsis*. Animals may display fur patterns that resemble their surroundings so they are difficult to see; for example, desert animals are often sand-coloured, and polar animals commonly have white fur. Alternatively, some animals, usually those that live in water, use transparency as a means of camouflage. Several species of fish – particularly at the juvenile stage of development – and many species of jellyfish are not just transparent, but their bodies match the refractive index of the water they live in, so no strange distortions are visible as they move through it.

High levels of transparency are, however, difficult to achieve and come at a cost. To give the appearance of transparency, cells in the body must contain a lot of water, and this means they are restricted in what else they can contain. For example, transparent muscles are much less efficient. Transparent animals tend to be either slow moving creatures like jellyfish or aren't completely transparent, and vital structures within the body remain pigmented and visible. An alternative method that some fish use is silvering effects on the skin that gives the illusion of transparency in the water. Transparency out of water is even more difficult to pull off because cells would have to have the same refractive properties as air to look convincing, and air is much less dense than water. Insect wings are virtually transparent because they are very thin.

The majority of these camouflage techniques tend to be permanent features for the animal in question. This would be of no use to a vampire that needs to change its appearance according to its immediate needs. But some animals can change their appearance as their surroundings or circumstances change.

Juvenile fish generally lose their transparency as they grow, as the benefits of being able to swim faster outweigh the need to hide. By contrast, animals such as the Arctic hare change their appearance throughout their lives. As temperatures rise, different genes are switched on inside the cells that produce pigment in the hare's fur. So, in summer they sport a greyish coat and in the winter, when the temperature drops, these genes aren't active, so the hare displays a white coat. Arctic hares that live in very northern regions, where summers are short and temperatures do not increase very much, have white coats all year round. It is a very simple and effective response to their environment.

Vampires need to change their appearance rather more dramatically and rapidly than an Arctic hare, but there are animals that can do this. Crab spiders, chameleons and cuttlefish can change their colouration over periods ranging from days to seconds. There are many different ways to change colour. Blood vessels below the skin's surface can dilate to make a human face flush red when embarrassed. Crab spiders secrete a yellowish pigment to almost disappear when they sit on yellow flowers. Chameleons have two layers of guanine nanocrystals within their skin, and by changing the spacing between these crystals, they reflect light differently, acting like thousands of tiny prisms. Their skin also contains chromatophores, cells full of yellow and blue pigments, that can be expanded or contracted to reveal or hide the colour. All contribute to the huge range of colours chameleons can display.

Cuttlefish take things a step further. They use their appearance as a form of social signaling, but also to blend into the background. Their colour changes are also controlled by a combination of chromatophores, containing different coloured pigments, as well as guanine crystals that can be manipulated to reflect light in different ways. But cuttlefish can undergo further transformations by altering the texture of their skin to look smooth or spiky to mimic smooth rock or spiky coral. This ability to mould the skin into different textures could possibly explain how vampires in *Buffy* and *From Dusk till Dawn* change their faces from smooth and human-like, to bumpy and demon-like when they attack.

The combination of colour and texture control in cuttlefish skin enables a staggering range of different appearances by the same animal. But even these transformative properties are not a patch on the champions of shapeshifting: octopuses.

7.3 FLEXIBLE FRIENDS

Octopuses can not only change colour, adopt different patterns – both static and moving waves or clouds of colour – by manipulating pigment-filled chromatophores, they can also alter the texture of their skin. This gives them a similar range of abilities as cuttlefish but, octopuses also have the advantage of their enormously flexible bodies, allowing them to make dramatic changes to their body shape, far beyond that of other creatures. As a result of this cosmetic complexity, some species of octopus have evolved into excellent mimics. The aptly named 'mimic octopus' (*Thaumoctopus mimicus*) has taken transformation and shapeshifting to impressively high levels. It has been observed to impersonate thirteen separate species, from lion fish to sea snakes, and jellyfish to anemones. If vampires had the physical traits of the mimic octopus, it would go a long way to explaining how they can transform themselves into different species, along with a few other characteristics of the blood-sucking undead.

The ability to rapidly change the appearance of the skin could allow vampires to display the desired shape – perhaps a bat – while the rest is camouflaged to blend into the background and give the appearance of invisibility. There would be no need for sudden gains or losses of body mass, it would just be an illusion. Having an octopus's ability to manipulate their body shape would make transformations more convincing. It could also potentially explain how the undead escape their coffins each night without disturbing the soil, or squeeze through the cracks of a mausoleum, to go about their blood-sucking business.

Changing shape is difficult, if not impossible for humans due to our rigid skeleton, but other animals with internal skeletons manage a far greater range of flexibility. Cats are incredibly supple due to stretchy ligaments, and can squeeze through small gaps. Mice, thanks to their flexible skeleton can fit through any gap bigger than their skull. But nothing compares to the octopus. With no bones to restrict it, an octopus's body can be squeezed and stretched into a fantastic range of shapes using muscular hydrostatics. The total volume of the octopus does not change, but its shape can vary enormously.

Muscle hydrostats are not just found in cephalopods. An elephant's trunk, or a human tongue can adopt a range of shapes without changing the overall volume. This is possible because the cells that make up an octopus's tentacle, or an elephants trunk, are filled with incompressible liquid. If the cells are squeezed in one direction, they must expand in another direction. A worm can shorten and lengthen its body to move, but each contraction makes the worm fatter and each stretch makes the worm thinner. A human tongue can be flattened, rounded and curled into a variety of shapes, but it doesn't get any bigger or smaller. Almost the entire body of an octopus is made up of muscular hydrostats and therefore the only limitation on the gap they can squeeze themselves through is the size of the hard beak they use to crush their food. National Geographic has shown that an adult octopus weighing nearly 300 kg can squeeze through a gap the size of a two-Euro coin, which are 25.75 mm in diameter, or roughly one inch.

Despite the fact that the animal that best describes the superhuman shapeshifting attributes of a vampire is an octopus, these animals are not normally associated with them.[‡] There is such a thing as a vampire squid (*Vampyroteuthis infernalis*, which literally translates as 'vampire squid from hell'), but this animal got its name because of its appearance, rather than any shapeshifting or blood-sucking traits (see Figure 7.1). Vampire squid vary in colour, from dark velvety black to pale red, they also have huge red eyes (the largest of any animal in proportion to their body) and webbing between their tentacles that resemble the opera capes so often worn by vampires in films. Under their

[‡]I am happy to be corrected if you know of any examples.

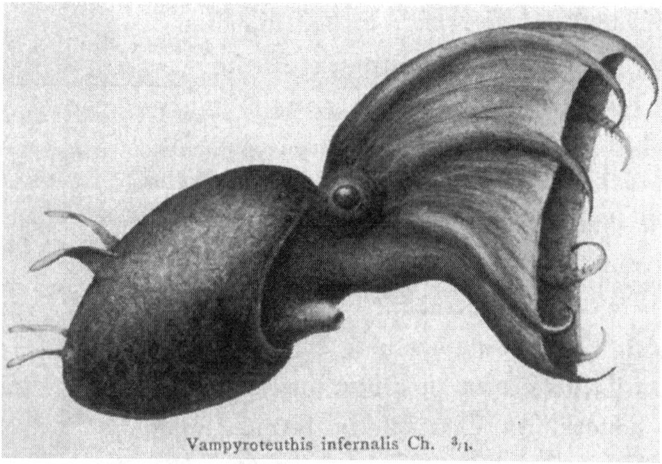

Figure 7.1 *Vampyroteuthis infernalis* illustration by Carl Chun, 1903.

cloak, their tentacles are lined with fleshy spikes that look at first glance like fangs or pointy teeth. They live in the depths of the ocean where virtually no light reaches, but they are not skulking in the shadows looking for blood. These animals feed on detritus, and are more likely to be food for bigger predators than the other way round.

Vampires might not make you think immediately of octopuses, but they are strongly associated with other animals. However, the animals they are normally thought to be able to transform themselves into, don't have the kind of flexibility or camouflaging talents that could explain a vampire's shapeshifting abilities. In which case, of all the animals that a vampire could turn into, why are they mostly associated with bats, dogs, wolves and rats?

7.4 WOLVES AND DOGS AND RATS, OH MY!

There is a particularly strong association between dogs or wolves and vampires. Dracula takes on the shape of a huge dog to leap from the ship that carried him to England. The Count also uses his control over wolves to borrow one from London Zoo and force it to break the glass in Lucy Westenra's bedroom window. In some traditions, vampires can control wolves and dogs to make them do their bidding (see Chapter 6), or they can even change their shape to appear as dogs or wolves, just as Dracula can. Stoker would have read about the links between vampires and these animals during his extensive research for his novel. One book he consulted, Sabine Baring-Gould's *The Book of Were-Wolves*, describes how in Greece, vampires and werewolves were considered to be closely related. In Serbia, the same word, *vlkoslak*, is applied to both vampires and werewolves. Baring-Gould also linked the werewolf's well-known taste for human flesh and blood with notorious stories of cannibalism and cruelty. His book includes a chapter on the famed 'Blood Countess', Erzsebet Báthory (who we met in Chapter 4) and his account may have inspired Stoker's three vampire brides that reside with the Count.

That dogs or wolves could be associated with the undead is not that surprising. These wild animals might well be attracted to graveyards because, in the past, bodies were not necessarily buried in coffins, and graves may not have been that deep. The smell of decomposing bodies could attract scavengers such as dogs and wolves. Wolves are fairly secretive creatures and would probably have visited graveyards at night to avoid contact with humans, living ones anyway. Their digging around a fresh grave, or perhaps even biting or trying to remove a corpse, could be interpreted very differently in the morning light. By then the wolf has gone and all that can be seen is disturbed earth around a recent burial and perhaps even evidence that the body has moved. If the wolf was spotted around the graveyard, this could have been interpreted either as the animal attacking the occupant of the grave or being the occupant of the grave himself, but in a different physical form. In other traditions, however, dogs and wolves are the sworn enemies of vampires and a white wolf in a graveyard was said to protect locals from vampire attacks.

Rats, like other scavengers, would be attracted to decaying bodies. Their presence in increased numbers during an outbreak of disease, that killed a number of people in a short space of time, would be understandable. Rats would also be a rational explanation for a particular type of undead that plagued certain parts of Europe and were said to chew on their own fingers, or the bodies in neighbouring graves. Equally understandable would be the presence of large numbers of flies and other animals that feature prominently in *Dracula*, though not so much in other vampire tales. In folklore tradition, the animals most commonly associated with blood-sucking revenants were birds, insects, rats, dogs, wolves and, in only a very few examples, were they ever linked to bats. The only real connection between bats and vampires in folklore is a vague reference to a bat flying over a corpse being able to transform it into a vampire.

Though bats are common in Europe, and are often associated with graveyards and deserted ruins, the European species have very little in common with European vampires. Vampire bats, on the other hand, are native to Central and Southern America. Terrifying stories of these blood-sucking creatures were brought to Europe with the return of Columbus and his crew in the fifteenth century. As bats operate in darkness, studying their behaviour is not easy, and early reports filled in any factual gaps with wild speculation, exaggeration and outright nonsense. Reports circulated of bats 'of such bigness' that 'assaulted men in the night in their sleep, and [have] so bitten them with their venomous teeth, that they have been ... compelled to flee from such places, as from ravenous harpies.' The general assumption was that these creatures were enormous, flying fiends that attacked humans and drained them of their life-blood. Quincey Morris in *Dracula* tells of his experiences in the Pampas, when his horse was attacked by a vampire bat and 'what with his gorge, and the vein left open, there wasn't enough blood left in her to stand up.'

As we discovered in Chapter 4, vampire bats are small, rarely attack humans and, though they have a healthy appetite, couldn't come even close to draining a horse or a human of its blood in a single night. But sixteenth-century stories of these bats, like vampires, refused to die. When, in the eighteenth century, Carl Linnaeus set about giving labels to all living things using his revolutionary system of taxonomy, any large, fearsome bat got the 'vampire' tag. *Vampyrum spectrum* (spectral vampire bat) is the largest bat in South America; it is carnivorous, feeding on rats, birds and other bats (including vampire bats), but it doesn't drink blood, and neither do the many other 'false vampires' that Linneaus named.

Already suffering from a bad press, things only got worse for bats after they featured so prominently in *Dracula*. Initial connections between bats and vampires were made through misunderstandings about a few blood-drinking species and far-fetched stories from exotic locations. But as time went on, bats became inextricably linked to vampires and more and more connections were found between the two species. The reputation of bats has suffered considerably as a result.

7.5 'THE BAT IS MY BROTHER'

Robert Bloch was a prolific writer of crime, fantasy, horror and science fiction short stories, as well as screenplays.[§] He wrote 'The Bat is My Brother', the tale of a vampire who wanted to take over the world by creating an army of fellow vampires. In Bloch's tale, vampirism is a disease that is spread through bites, and though he only mentions bats in the title of the story, bats have become closely linked to the transmission of several diseases.

A bat's ability to tolerate serious infections is just another of its amazing characteristics. Quite apart from their phenomenal adaptability, ability to fly and navigate by echolocation, which make bats so extraordinary, they also have the most incredible immune system.

Bats can fend off the most devastating attacks from bacteria and viruses that are quickly fatal to humans. It's not that the bats are invincible, but that their immune system will keep them flying and behaving normally almost until the point of death. When humans would be completely incapacitated by the same virus, these tiny mammals are still going about their daily, or nightly, business. It is yet another reason why bats make such an interesting subject for scientific research, as they may help us find ways to improve the response of the human immune system. Such an advanced ability to manage an infection also means bats are the subject of research for another reason. These animals can often be the natural reservoir for particularly nasty diseases. In different parts of the world, bats can be natural reservoirs for diseases such as Ebola, and it is possible

---

[§]He is best known for writing *Psycho*, which formed the basis for the Hitchcock film.

SARAS-CoV-2, the virus that causes Covid-19, was present in bats before it jumped to humans *via* another intermediary animal.

Bats can potentially spread these horrible infections to other animals through bites or by shedding the virus in their urine and faeces. However, given the number of bats in the world and how many hours bat researchers spend in close proximity to wild bats, transmission of these diseases directly to humans is a very rare event. Ebola outbreaks in humans, for example, have occurred in times and in regions where people are living under extreme socio-economic pressures that force them into places and to eat foods they would not normally consider. While it is true that any bat, not just a vampire bat, that is disturbed or picked up by a human may well bite to try and escape the situation, this is very much exceptional behaviour. Bats aren't really interested in humans and are best left to their own devices.[¶]

The fact that bats have the potential to spread disease by biting has given them another link to vampirism, where the most well known way of becoming a vampire is to be bitten by one. Even in folklore, vampires are associated with the spread of disease and local epidemics. The similarities between the symptoms of one disease in particular, and the characteristics of vampires, led one Spanish scientist to develop a new theory of the origins of vampires, and their association with bats.

7.6 RABIES

Rabies is a virus, a set of RNA instructions wrapped up in a protein coat, that kills between 50 000 and 70 000 people globally every year. Viruses, unlike bacteria, cannot replicate themselves and rely on their host to do the work for them. Inside an animal or plant cell is all the biological machinery needed to make more cells so it can repair itself and replicate. Viruses insert themselves into cells, where they hijack the growth and repair machinery and use it to make more viruses. This process continues until the cell literally bursts apart, releasing thousands more viruses into the system to infect more cells.

[¶] Bats are protected species. If you find a bat, please don't pick it up, you are likely to make it more stressed. Call the RSPCA or a similar organisation with the expertise to deal with bats safely.

The rabies virus is transmitted by bites from infected animals. Viruses in the saliva enter the wound and initially multiply in the surrounding muscle tissue. Then they enter the nerves and slowly travel along the network of cells that make up the nervous system. During this initial phase of infection and migration, there may be few symptoms. There might be some itching around the original bite, flu-like symptoms or weakness and headaches, but nothing specific or very distressing. This can last for up to ten days, but then disappears while the virus travels ever onwards to its target destination, the central nervous system (the brain and spinal cord).

How long it takes to reach the central nervous system depends on several factors, including the strain of rabies virus and the distance between the bite and the brain. The result is huge variation in the incubation period of between a few days or several months, but it is usually two weeks to two months. Bites to the neck or chest are more alarming than bites to the hands or feet. Children are particularly vulnerable to bites from rabid animals because they are physically smaller.

Bats, with their tiny bodies, would be expected to quickly succumb to the effects of rabies, but they continue to fly about apparently unaffected for months with the virus inside them. One theory is that the high metabolic rate bats need in order to fly also gives them an increased ability to fight off the effects of an infection. It is only in the very last stages of the disease, when the bat is close to death, that its behaviour might change, for example, flying in daylight.

In humans, and other mammals that aren't bats, once the virus arrives at the central nervous system, a new set of very obvious and far more frightening symptoms appear. In around 20% of people there will be a slow flaccid paralysis that results in quadriplegia, sometimes known as 'dumb' rabies. More common is 'furious' rabies, characterised by hyperactivity – restlessness, a feeling of terror, persistent insomnia, increasing agitation and a tendency to delirium. A few days after the onset of these behavioural changes, a new symptom appears, one that is characteristic of the disease – hydrophobia, or fear of water.

The symptom of hydrophobia is unique to rabies. Despite a raging thirst, attempts to drink result in an overactive reflex in the throat that produces a gagging sensation. This gets progressively worse until the sight, sound or even thought of water becomes incredibly distressing. These episodes can last up to five minutes.

The muscular spasms that accompany the hydrophobia stage can mean the face is twisted and the lips pulled back into an aggressive snarl. Saliva can't be swallowed because of the gag reflex and so there is foaming at the mouth, which can be blood-flecked. There can also be vomiting of blood. These violent episodes can be triggered by the slightest thing, from draughts, to the sight of water, bright lights, loud noises, strong smells or even the sight of the victim's own reflection in a mirror. In the past, when medical knowledge was considerably less sophisticated than it is today, rabies was sometimes diagnosed if a person could not look at their own image in a mirror.

After two to seven days of disturbed behaviour, furious rabies ends with coma or sudden death from respiratory or cardiac arrest. For the paralytic form of rabies, progression is slower and people can live up to thirty days before coma, lasting three to seven days, and death. In almost every case where the later symptoms present themselves, the result is death. Only a handful of people have survived the latter stages of rabies by being put into an induced coma and being aggressively treated with drugs, but the damage to the nervous system almost always leaves them severely physically and mentally disabled.[‖]

Rabies can infect any mammal, but it is most commonly dogs, foxes and wolves that transmit the disease to humans. Bats have been known to pass on rabies to people through bites but the number, compared to transmissions between dogs and humans, is tiny. Human-to-human transmission is rarer still, and the few cases that have occurred have been through sexual contact, bites or when nursing patients.

The first person to link rabies and vampires was a Spanish neurologist named Juan Gómez-Alonso. He had gone to the cinema and seen his first vampire film in 1981. What he saw depicted on the screen struck him as remarkably similar to the rabies victims he had treated professionally. It sparked an

---

‖ Rabies vaccines are extremely effective for people who might come into contact with rabid animals. Anyone who is bitten by a rabid animal should get treated as quickly as possible. Recovery is usually excellent with swift medical intervention.

interest that led him to investigate vampire stories more deeply and he published his theory, 'Rabies: A Possible Explanation for the Vampire Legend', in 1998.

There are certainly many similarities between the symptoms of rabies and the symptoms of vampirism. The biting, aggression, insomnia, aversion to bright lights and strong smells such as garlic, and the fear of water and mirrors all tally well with characteristics commonly associated with vampires. The link to certain animals often associated with vampires is also strongly in its favour as an explanation for the phenomenon of vampires.

In further support of his theory, Gómez-Alonso also pointed out that there was a huge outbreak of rabies amongst wolves and other wild animals in Hungary around 1721–1728. The dates, if not the exact location, coincide with the most famous vampire accounts of Arnold Paole and Peter Plogojowitz that sparked the Europe-wide obsession with vampires (see Chapters 2 and 3). In the past, when nothing was known about bacteria or viruses, the true cause of the disease was impossible to determine. Perhaps, seeing several people die in quick succession as a result of attacks from rabid dogs or wolves, a link to stories of revenants that could transform into these animals would not have been hard to make. In some traditions, the neurologist claimed, the vampire's lifespan was not eternal, as in the films, but limited to 40 days – a not unreasonable period for the development of rabies in an individual. But there are problems with the theory.

The vampires of European folklore were most definitely dead before they came back to prey on the living, not sick for several weeks and then dead. The association of blood-sucking revenants with wolves and dogs is strong in both traditional folklore and fiction, but bats stem almost entirely from the novel *Dracula*, published long after these major rabies outbreaks but at a time when the disease was still a concern throughout Europe.

However, the situation in the Americas in the early twentieth century, was different. There they have vampire bats and, unlike their European counterparts, they bite other mammals all the time – it's what they are good at. However, vampire bats only really became a nuisance with the expansion of cattle farming in this part of the world. They are often seen as a pest that annoys cattle, or occasionally a person who pokes their toe outside the bedcovers at night, but they are not seen as a particular threat or danger. Only a small percentage of the bat population is normally infected with rabies, estimates are around 6%, but occasionally the effects on other animals can be devastating.

7.7 THE RETURN OF THE *SOUCOUYANT*

In Trinidad in 1925, there was an outbreak of mysterious disease in cattle. The animals would have saliva dripping from their mouths, they staggered and eventually became paralysed, but no one knew why. Searches were made for poisonous plants growing in the fields, but nothing was found. Botulism was suspected, but tests proved negative.

In 1929 things got worse. One thousand cattle died and the disease spread to horses, dogs and humans. Fifteen people in the town of Siparia became paralysed and all of them died. In rural areas some thought *soucouyants* were the cause of it all, from their preying on people and livestock. Despite many theories as to the cause of the outbreak, no scientific evidence could be found to back them up. That was until a farmer saw a vampire bat clinging to the flank of one of his herd. A week later, the cow was dead. Tests proved that the cause of death was rabies, but it was only speculation that it was the bat that had transmitted the disease to the cow. Research into the island's 32 different bat species was almost non-existent at the time.

Then, one day, a boy working on a farm heard some squeaking. He looked up into a palm tree to see two bats fighting each other. It was an extraordinary thing, bats of the same species do not normally fight each other and wouldn't be expected to be active in broad daylight. When the bats fell to the ground, the boy killed them with a shovel. Knowing about the increased interest in bats, the boy took the squashed bodies to a scientist, Dr Metivier. Metivier injected an emulsion made from the bats' saliva into a young calf, a guinea pig and a rabbit. All three animals died. It seemed that bats really were the cause of the strange illness spreading through the island.

A great bat hunt began. Every nook and cranny was searched, and 88 bats were brought in for scientists to examine. They found eight of these bats had the tell-tale Negri bodies in their brains. These Negri bodies are small red spots found in the brains of rabies victims, named after their discoverer, an Italian pathologist Adelchi Negri. It confirmed that the bats were infected with rabies and could potentially pass it on to other animals through bites. Negri bodies were found in only two species of bat, the fruit bat *Artibeus palmarum*, and the vampire bat. The little fruit bat was probably only a secondary carrier, as it usually only sank its teeth into fruit.

People do not always take the trouble to distinguish different bat species and assume all bats are a menace and should be culled. However, bats are essential for healthy ecosystems. Many species of bats are vital for agriculture because they pollinate certain plants. Other bat species are beneficial because they eat the insects that would otherwise damage crops. Rather than wholesale slaughter of bats, the most obvious way to prevent the spread of rabies was to stop them biting domesticated animals and humans. This was relatively easy to do for humans, who were rarely the victims of bites in any case. Screens at windows, or netting over beds was all that was really necessary. But bat-proofing the sheds where cattle were kept overnight was beyond the means of many farmers. An alternative was to leave bright lights on to deter the bats, but only strong electric lights worked, and most of the island was not on the electrical grid at the time.

The only way of effectively halting the disease in its progress was through vaccination. An effective vaccination had been developed in 1885 by Louis Pasteur, which could be used against all forms of rabies, in dogs, cattle and humans. Vaccinations started with cattle at the farmer's own expense, but in 1933, a campaign was started to vaccinate cattle and to urge any people bitten by bats to seek urgent medical attention. Slowly, the epidemic was brought under control, but not before many farmers had been left destitute by the loss of their livestock. Overall, 5000 cattle and forty people had died, but it was the last major outbreak of the disease. Rabies is still endemic in Trinidad, and small outbreaks continued to occur until the mid-twentieth century. And Trinidad is not alone in its bat-related rabies problem.

In 2008, a rabies outbreak in Venezuela was linked to vampire bats. The disease killed 38 people, but was swiftly brought under control very simply, by distributing mosquito nets. There are also a few cases of rabies reported each year in the United States, and are mostly due to bites from bats (though not necessarily vampire bats).**

By the end of the twentieth century, with increasing access to electric lights to deter the bats, and with fewer mortars being used, preventing their use as storage for human skin, the *soucouyant* had become an endangered species. Elsewhere, the shapeshifting abilities of vampires, and their close association with bats, had become firmly established. But it had little to do with folklore traditions of vampires or stories of vampire epidemics in Europe in the eighteenth century. The reason these revenants acquired these extraordinary shapeshifting abilities probably has more to do with aesthetics.

Dramatic transformations into bats can make for stunning special effects in films. Capes can be cut to give impressive bat-like silhouettes on stage and screen. The association of bats with darkness and deserted buildings makes a useful shorthand in vampire fiction. Many sinister moments can be created in novels like *Dracula*, where the vampire can be simultaneously disguised from the other characters in the story, but also revealed to the audience through their ability to change into another animal associated with vampirism.

Though rabies has been known to infect dogs and wolves for centuries, the link between bats and the spread of disease came about in the twentieth century. This association between bats and contagion tied in with the already well-established links that had already been made between vampires and lethal illnesses. Rabies is not the only communicable disease that has been linked to vampires, but its symptoms and method of transmission make it a very convincing explanation for many modern vampire characteristics. Could the vampire epidemics of eighteenth-century Europe also have been due to diseases with vampiric symptoms? That is the subject of the next chapter.

**Elsewhere in the world, dogs and foxes are the major transmitters of rabies infection.

CHAPTER 8

Disease

> 'I must keep aloof from everybody, that I am a social leper, full of contagion and death!'
>
> Florence Marryat, *The Blood of the Vampire*

It was the middle of the night of 17 March 1892, and a small crowd had gathered around the final resting place of Mercy Brown. Her tomb was opened and her decaying heart removed from her chest before it was burnt. The ashes were carefully gathered up, mixed with water and given to Mercy's younger brother, Edwin to drink. He had been ill for months, weak and pale, and now it was clear he was dying. This was a last-ditch attempt to save his life.

The Brown family of Rhode Island, New England, had been hit hard by a series of deaths in previous years. It had all started in December 1883, when Mary Eliza, the mother of the Brown family, was buried in Chestnut Hill cemetery in Rhode Island, New England. Seven months later, her daughter Mary Olive followed her mother to the grave. It was a few years after these first two deaths that Edwin first became sick.

Edwin lost his strength and appetite, and his complexion became pale. He was sent to Colorado, hoping for a cure, or at least that the progress of his illness could be slowed. But the move did not work. His health continued to deteriorate and it was decided he and his wife, who had accompanied him to care for him,

Vampirology: The Science of Horror's Most Famous Fiend
By Kathryn Harkup

Published by the Royal Society of Chemistry, www.rsc.org

would return to their native New England to see out his final days. During Edwin's absence, Mercy, another sibling in the Brown family had also fallen ill. She died before Edwin could reach home.

Mercy was just nineteen years old when she was placed in the family vault in January 1892. The father of the family, George Brown, having buried his wife and two daughters, was now watching his son Edwin die. Many in the community urged George to do something to save his son. But what could be done when the medical treatment in Colorado had failed?

There was one traditional method used to stop the curse that appeared to have fallen on the family, and members of the local community sought Mr Brown's permission to carry it out. Not believing in the folk remedy, but perhaps desperate to try anything that might save his son, and maybe a little out of fear of upsetting the strongly held beliefs of the community, Brown senior consented for the remedy to be tried. This traditional method was based on the belief held by some of George's friends and neighbours that Mercy was the cause of the Brown family's terrible situation. She was, therefore, also the cure.

On 17 March, a group of people, including the local physician Dr Metcalf, went to Mercy's grave. The doctor was reluctant to be involved in the procedure but consented to carry out an autopsy on Mercy's body. When he arrived at the cemetery he found that four men had already removed the remains from the vault where she had been placed. They reported that Mercy's body had moved during her time there. Though originally she had been laid out in traditional style, on her back with her arms folded across her chest, two months later she was found to have rolled onto her side. They also noted that her body was remarkably well preserved.

There was nothing unusual about Mercy's appearance when she was exhumed in March. As we know from Chapter 3, bodies move and burials in colder months are likely to preserve remains better than those in summer months. Mercy's body had been placed in the vault precisely because the ground had been too frozen to bury her when she had died two months earlier. But, the witnesses took these signs as unnatural. Mercy, they thought, was restless, and sustaining herself after death by preying on her brother.

The doctor proceeded with the autopsy in less than ideal circumstances. He opened Mercy's chest and examined her heart, where he found liquid blood inside. He assured those present that nothing he had seen was at all unusual, but those around him were not convinced. They believed the blood was not the clotted fluid that normally occurs as part of natural decay, but that it was blood taken from Mercy's brother and this was the cause of his wasting illness. They took the heart, burnt it to ashes on a nearby rock and prepared their remedy for Edwin.

You may not be surprised to learn that Edwin Brown was not cured of his illness. He sadly died two months later. Stories of these strange events soon spread. Gossip circulated and articles were written in newspapers about the New England vampire that had come back from the dead to feed on the life of her brother.

Mercy Brown has become, since her death, perhaps the most famous American real-life vampire, but the vampire tag was only given to her by others. Those in New England; the family; those involved in opening her chest and burning her heart; the community that insisted on the procedure, none of them used the word 'vampire'. It was others who read reports of this folk remedy and interpreted it alongside their existing knowledge of stories like 'The Vampyre' and *Carmilla.* Newspaper reports gave Mercy the vampire label and she has been stuck with it ever since.

The cause of Edwin Brown's death was consumption, or tuberculosis, not the vampiric predations of his sister. The same disease had killed Mercy, her sister and their mother. During the doctor's examination of Mercy's body, he had seen that the lungs contained clear signs of the disease. What the community did was, at that time and in that location, a well-known way of dealing with the terrifying threat of consumption, a disease that spread and killed apparently indiscriminately and uncontrollably.

8.1 CONSUMPTION

Tuberculosis (TB) has been with us for thousands of years. A background of continuous infections and deaths from the disease have been punctuated with occasional epidemics and periods of especially high mortality. The United States saw an alarming increase in TB cases from the 1730s. The eastern United States were particularly badly affected, and by 1800, it accounted for a quarter of all deaths and remained the leading cause of death well into the twentieth century.[†]

TB is a bacterial infection, usually of the lungs (pulmonary TB), but it can infect any part of the body (extra-pulmonary TB). The bacteria, *Mycobacterium tuberculosis humanis*, was identified in 1882 but the name tuberculosis first appeared in 1840. It derived from a Latin word for swelling because of the small nodules, or tubercles, that appear in the tissues of those infected. However, it was widely known as consumption because, no matter how much an individual ate, they appeared as though they were being consumed from the inside.

The disease causes night sweats, loss of appetite, weakness, fever, pain in the sides or at the shoulder. Some historical accounts speak of the breath of a consumptive having a tar-like odour.[‡] Over time, the tiny tubercles in the lungs develop into abscesses and cavities that produce an abundance of sputum and give rise to a consumptive's hacking cough as they try to clear the fluid from their lungs. The lesions that form bleed and then haemorrhage. The mucous starts off blood-flecked, and as the disease progresses, blood loss increases from teaspoonfuls to cupfuls in the latter stages. This loss of blood then causes anaemia.

The symptoms are worst at night. Lying asleep horizontally, fluid accumulates in the lungs and it can feel like a huge weight is pressing down on the chest, or like someone is sitting on you. The untreated victim of TB will waste away, leaving them with a pale, death-like pallor. It is not so very different from the reports from the vampire victims of the 1720s who complained of revenants who sucked the breath from them or suffocated them in their sleep, leaving them wasted and weak by the next morning. In Rymer's epic vampire saga *Varney the Vampire*, he makes specific reference to the disease: 'And that the hideous repast of blood has to be taken frequently, and if the vampyre gets it not he wastes away, presenting the appearance of one in the last stage of consumption, and visibly, so to speak, dying.'

[†]TB still kills between 1.5 and 2 million people every year. It is the number one killer in terms of infectious disease.

[‡]Perhaps a link to the foul stench complained of on Dracula's breath.

TB is transmitted through droplets coughed into the air and inhaled by others close by. In the nineteenth century, those living in cramped homes with poor diets would have been most susceptible. How an individual responds to a TB infection also depends, in part, on their genetic makeup. In the past, consumption appeared to run in the family, just as the Brown family in Rhode Island shows. Not only did they live together, giving the best opportunity to transmit the disease between different members of the family, but their close genetic relationship meant they were likely to have a similar susceptibility to infection. Folkloric vampires were known to prey on their nearest and dearest first, followed by neighbours and close friends, a pattern that can easily be explained by the spread of an infectious disease like TB.

In Mercy Brown's case, she had been diagnosed with the 'galloping' variety of TB. Galloping consumption was used to describe the situation where symptoms appeared suddenly and there was a rapid decline in health. In fact, the disease would have been with her for a long time before then, but unrecognised and undiagnosed. TB can enter a latent state, where symptoms disappear; the bacteria are still there but lying dormant inside the lungs waiting for an opportunity when the human immune system weakens and they can re-emerge and multiply. It was possible to live decades of normal healthy life up until the disease made a dramatic re-appearance and seemed to kill its victims in a matter of months or weeks.

Though the bacteria had been identified by the time the Brown family became sick with the disease, there was no effective cure available. The best that could be done before sulfonamide drugs were available – the first antibiotics that were developed in the 1930s and 1940s – was to isolate TB patients in sanatoriums. It separated the sick from the well and prevented the spread of the disease. While in the sanatorium, rest and an improved diet gave the body the best chance of fighting off the infection.[§] Even with dedicated long-term care and modern drugs, however, those that survive TB are often left with

[§]Today, TB can be treated with antibiotics, but some strains of the bacteria have become resistant to drug therapies. It can take weeks or months of treatment with drugs that produce unpleasant side effects to completely eliminate these resistant strains.

permanently damaged lungs. The prospects for TB patients in the past were, therefore, grim indeed.

The similarities between consumption symptoms and its spread, and the characteristics of vampire activity, made it an easy connection to make. In the late nineteenth century there was no effective treatment for the highly contagious and deadly disease. The appearance of symptoms in one person signalled that other family members were likely to catch it and from there it could spread to the wider community. Without effective antibiotic treatment, TB kills around half of those it infects. Even if you survive an infection, it does not confer immunity and you are actually *more* likely to catch it again. People in the nineteenth century and earlier were understandably frightened and went to extraordinary lengths to try and protect themselves from a devastating disease. Mercy was not the first person in New England to be disinterred and her body mutilated in an attempt to stave off an epidemic.

In 1990, three ten-year-old boys were playing in a gravel pit in Griswold, Connecticut where they found two human skulls. A spate of killings that had occurred in the area a few years before initially had people worried that the boys had discovered more of the serial killer's victims, but the bones were far too old. The state archaeologist was brought in and he quickly pointed to where the gravel pit had cut through an old graveyard. Further excavations of the burial ground revealed a mysterious stone crypt. Inside were the bodies of a mother, father and their son, but there was more to this burial than a family tragedy. The father's remains had been tampered with; the skull had been turned backwards, to face away from the rest of the body; the leg bones had been placed in a cross position, pirate-style, over the rib cage, and someone had been rummaging around in the chest. Closer examination of the ribs showed the tell-tail lesions of tuberculosis.

There is evidence of a dozen such cases in this region of the United States, but Mercy Brown was the last and certainly the most famous. People still go to her tombstone, especially on Halloween night, to see the grave of New England's vampire. Chips have been taken from her gravestone as souvenirs, and on one occasion the whole stone was removed, though it was returned less than a week later. Her headstone is now secured to the ground with a heavy iron collar. She receives all this attention even though she was just a young woman who died of a terrible disease and whose family and community were desperate to save others from the same fate. The community was rural and isolated. There was little access to medical help, not that medical science could have done much for the Brown family in any case. People had to do what they could for themselves.

TB shares many of the characteristics of both folkloric and modern vampires. Outbreaks of the disease in rural populations would have led to several deaths occurring in rapid succession. The sudden spate of deaths was probably not as rapid as the reports of the well-known vampires Arnold Paole or Peter Plogojowitz would suggest (see Chapters 2 and 3), though there are many other parallels that can be made, particularly in how locals responded to the deaths. The pale, gaunt appearance of the consumptive was fashionable during the early nineteenth century and it is no coincidence that vampires took on this look when they became the subject of romantic literature (see Chapter 10).

Consumption wasn't the only disease claiming lives in the nineteenth century, though it certainly led the way in terms of the huge number of victims. There were plenty of other contagious diseases that killed Victorians, several of them also with links to vampirism. One of the most terrifying, in terms of its sudden appearance and swift and devastating mortality, was cholera.

8.2 CHOLERA

The name 'cholera' has a long and confusing history, which is appropriate for a disease that still defies strict definitions. Much of the confusion can be traced back to the four humours theory of medicine, introduced back in Chapter 4, which held sway over medical thinking for almost two thousand years and lingers on in our everyday language. The four humours were blood, phlegm, black bile and yellow bile. Choler, a word meaning irritability or anger, was thought to be caused by an excess of yellow bile. When the expulsion of excess yellow bile became severe enough to be considered an illness, that illness was called cholera. Therefore, before the nineteenth century, cholera was associated with almost any kind of upset in the stomach or guts that produced vomiting and diarrhoea. Such illnesses could be mild or deadly, but in 1817, cholera became something much, much worse.

An outbreak of cholera in August 1817, retrospectively labelled 'the first pandemic', occurred in Jessore in the Ganges delta. This was no ordinary upset stomach, it killed 10 000 people – half those it infected – and unlike previous instances of cholera, it spread. By September, the disease had moved to Kolkatta, and by November it was a thousand miles away in Allahabad, where thousands died in a matter of days. The second pandemic, that began in 1829, went global. By 1830, it was in Moscow; by 1831, it was in London; it was in Paris by 1832, when it also crossed the Atlantic to arrive in the United States.

People sickened and died within 24 hours. Fluid streamed uncontrollably from the body, causing the patient to appear to shrink and age in the space of a day. Cholera victims withered away in front of their family's eyes. Muscle spasms and cramps contorted the body. The skin turned blue, the blood thickened and turned black. Many would fall into a coma. Mass graves were dug, and some, out of fear of contagion, were thrown into the pits before their time. Some revived to find themselves surrounded by dead bodies.

Charlotte Stoker told her young son Abraham about the time cholera had ravaged her home county of Sligo in Ireland. Friends and neighbours who had been hale and hearty the day before became wizened and skeletally thin overnight. The young Stoker, bedridden by a mysterious illness, was enthralled by his mother's tales.

For decades, no one knew the cause of the new cholera, as it was sometimes called, or how it could be transmitted from one person to another, but many speculated. There were theories of miasmas, *i.e.* bad smells; eating too much green fruit; changes in the weather, or combinations of these many different factors. The celebrated story of John Snow removing the handle of the Broad Street pump in London in 1854, and thereby dramatically cutting the number of cholera cases in the area, showed that the disease could come from contaminated water (see Figure 8.1). He is credited as the father of epidemiology – the study of cause, distribution and control of disease – but the data he gathered about the pump and the cholera cases that sprang up around it didn't explain everything about the disease (see Figure 8.2).

Figure 8.1 Nineteenth-century cartoon by George Pinwell highlighting John Snow's recent discovery of cholera in the water supply, 1866.

In 1884, Robert Koch identified the comma bacillus, now named *Vibrio cholera pacini* in honour of Fillipo Pacini, the Italian scientist who had isolated the bacteria decades earlier but whose work went unrecognised until much later. Finding the causal agent still didn't answer every question about the disease. Some could have the bacteria in their gut, but show no symptoms. The more people looked, the more places the bacteria were found, and often apparently causing no harm at all.

The bacteria itself did not cause the symptoms of cholera, it was a toxin released by the bacteria. Different strains of the bacteria may or may not have the genetic code to make the toxin in the first place. Cholera exchanges DNA so easily that

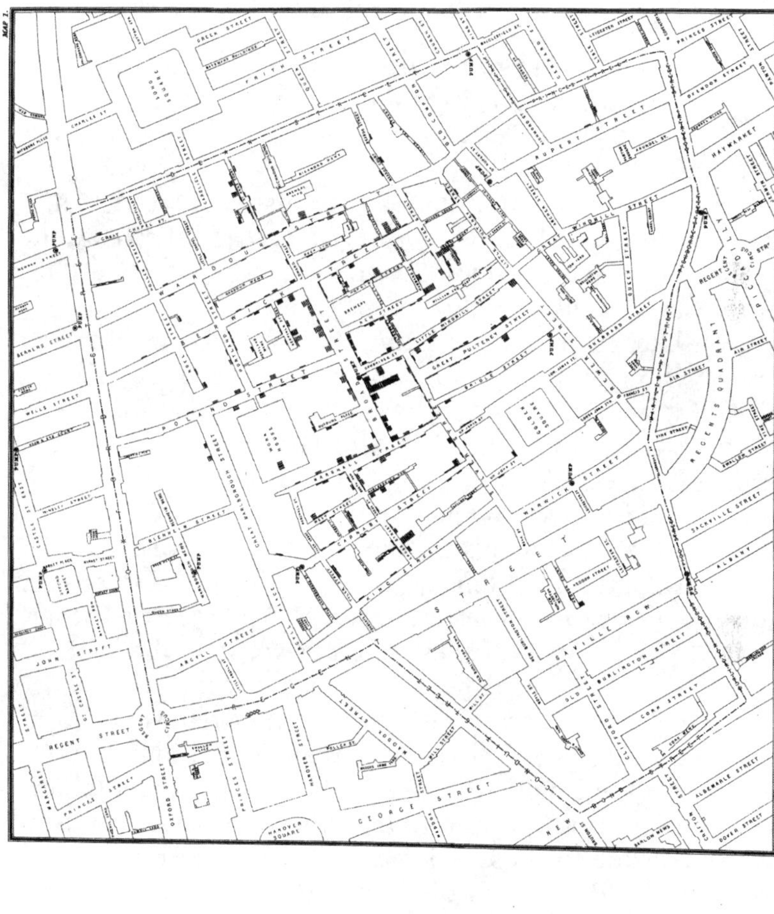

Figure 8.2 Original map by John Snow showing the clusters of cholera cases in the London epidemic of 1854, drawn and lithographed by Charles Cheffins.

classifying different strains can become irrelevant. In addition, a whole series of factors needs to be in place for this gene to be switched on and for the toxin to be released to cause problems in the gut. Cholera is normally a mild-mannered marine organism, but it can be transformed into a lethal human pathogen because of environmental changes or DNA exchanges, prompting it to behave differently. Some of the lethality of cholera comes about because it suddenly finds itself trying to survive in a human gut, rather than in sea water.

If the bacteria moves from a marine environment to a human intestine, it can produce a compound that helps clear away the mucous lining the gut walls to make conditions more amenable to the bacteria. This compound also has the unfortunate side-effect of producing the symptoms of the disease cholera. This molecule, or toxin, doesn't itself directly damage the body. It tricks cells into releasing a flood of water and salts, which results in the rice-water stools that characterise cholera.

The toxin latches on to cells lining the gut and once there it interferes with the normal operations of cyclic adenosine monophosphate (cAMP). Among a variety of important roles, cAMP controls the movement of ions (dissolved salts) across cell membranes. It does this by binding to ion channels that funnel potassium, sodium, calcium and chloride in and out of cells and water molecules flow alongside these ions. Ions need to be concentrated or excluded from certain places in order for cells to function properly. What is normally a well-regulated cycle that keeps everything on an even keel breaks down when the cholera toxin gets involved. It becomes a non-stop process, emptying salts and water into the intestines creating the extreme diarrhoea and severe side-effects associated with massive salt loss.

The simplest, and most effective way to treat cholera is not to neutralise the toxin, or even to try and kill the bacteria, but to compensate for the losses it causes. Drinking water with potassium and sodium salts dissolved in an appropriate concentration is very effective. A little sugar also helps the medicine go down because it makes it easier for the gut to absorb the sodium.[¶] If vomiting prevents liquids being drunk, then intra-venous (IV) drips can be used. But IV drips are a

[¶]Any source of sugar or starch will do, from rice to refined glucose.

twentieth-century invention. Even injections were not as common in nineteenth century as you might think.

Instead, people tried all sorts of treatments to cure, or at least alleviate the suffering, of those with cholera, and only a few noticed the benefits of drinking water or even injecting it. R. Nelson, writing about cholera in 1866, commented on the miraculous revival of a patient, 'The moribund state of this strong young man before the injection and his sudden apparent recovery gave to many of the bystanders the idea of a ghostly resurrection of a corpse; some of them left the room in terror.'

The nineteenth century produced a lot of cholera literature, from theories of its origins and spread, to methods of treatment. In amongst the many pamphlets, books and tracts was all the necessary information to control and treat cholera effectively, but it got lost in the volume of material and deadly pandemics continued to wreak havoc on populations until the early twentieth century. Outbreaks still occur today, but mortality rates have plummeted with effective treatment and containment.

Cholera, and the epidemics it caused in the nineteenth century, have a lot in common with vampire characteristics from folklore. A sudden spate of deaths in a community, the rapid deterioration in health and appearance of those affected, sounds very similar to the Arnold Paole and Peter Plogojowitz cases. Possible revivals after premature burial or simply drinking water, have the dramatic appearance associated with the undead rising from the grave. A mix of characteristics from TB and cholera were evidently selectively incorporated into the vampire literature of the Romantic era and Victorian fiction. And there was yet another disease that writers in the past could have used for inspiring terror in their predatory blood-sucking anti-heroes.

8.3 THE POX

Different diseases provoke very different reactions from the public. In the early nineteenth century, TB was romanticised, cholera was associated with dirt and poverty, and syphilis was considered shameful. Syphilis, known more commonly at the time as 'the pox', is a sexually transmitted disease that, untreated, lingers in the body for years or decades, doing all kinds of damage before it finally kills its host.

The spirochete bacteria responsible for syphilis enters the body through mucus membranes or breaks in the skin and is therefore most commonly transmitted from one infected person to another through sex. Around three weeks later, the first visible sign of infection appears – a sore on the genitals. The sore can heal but the infection has not gone away, 'the outwardly serene surface' can disguise a 'mild and exquisitely chronic' progressive inflammatory reaction.

A sore on the genitals would deter many sexual partners, but once that has gone, the infected individual gives no sign that they harbour the disease, enabling its spread to new hosts. The next stage of infection is a rash; the body becomes covered in tiny pustules that will again disappear from view, but leave the patient with a whole host of new symptoms including aches and pains, loss of appetite and insomnia. Most people seek treatment at the first sign of the sore, and with modern antibiotics, that is usually the end of the matter. It was not so in the past.

Though many in pre-antibiotic eras sought treatment, these treatments were ineffective. However, disconnected periods of good health and illness could mislead an individual into thinking they had overcome the pox. In the years that followed, the original syphilitic sore was probably long forgotten. The range of symptoms produced by the bacteria's ongoing presence and the body's inflammatory response were attributed to a whole host of different maladies. Traipsing from doctor to doctor, diagnoses from rheumatism, to gout, eczema, headache, epilepsy, or just plain 'nerves', might be attributed to the ongoing poor health of the individual. Years or decades later, the disease reached its final stages – when syphilis affected the brain and aorta, the main blood vessel leading to the heart. It caused madness, an unsteady gait, and a weakening of the aorta's walls. The end finally came either slowly, with the person bedridden and raving about god, or suddenly, when the aorta gave way and massive internal haemorrhage killed them.

The symptoms of syphilis are so varied that it became known as the 'great imitator' and it would be easy to pick and choose amongst them to find similarities with the characteristics of vampirism. But that is not necessarily why this particular disease has often been associated with the blood-sucking undead and the novels written about them in the nineteenth century. It has more to do with the method of transmission. The pox was always a secretive and shameful disease because of its association with illicit sex. It is easy to draw parallels with husbands or wives sneaking off to meet a lover and the nightly visits of a vampire. A vampire biting their victim is a very intimate act and the contagion they passed on was a result of mixing their blood.

When John Polidori introduced the new breed of fictional vampires in 1821, they had noticeably greater sex appeal than their folkloric predecessors. The allusions to sex no doubt helped to sell more copies to the outwardly staid Victorian readers, and it became a major feature of vampire fiction. Veiled, and not so veiled, references to extra-marital affairs and infidelity are particularly strong in *Dracula*. The Count and Mina Harker were found in an embrace, with bare chests and blood everywhere, while Mina's husband was sound asleep. Lucy Westenra, who received blood transfusions from four different men, none of them related to her by blood or marriage, could be interpreted as the wanton woman whose true evil nature was revealed after she returned from the dead. The image of a sharp stake being driven through Lucy's chest by her former lover is a metaphor that hardly needs explaining.

Syphilis sufferers in the Victorian era were publicly shunned. People did their best to hide their infections and looked for discretion when consulting physicians and undergoing treatments. The most common remedy taken for the disease at the time was mercury compounds. The toxic properties of the metal poisoned the person as well as the parasitic bacteria living inside them. Mercury caused excess salivation, blackened teeth and damaged nerves. The huge volumes of saliva produced by mercury poisoning were part of the body's attempts to get rid of it. Although the saliva was outside the body proper, swallowing saliva only reintroduced it into the gut where it could be reabsorbed in a miserable cycle of poisoning and re-poisoning. The increased amount of mercury in the mouth irritated the gums, loosened the teeth and caused bad breath.

Mercury, with its strong affinity for sulphur, attaches to the sulphur atoms in proteins and enzymes, disrupting their function. Practically every protein and enzyme in the body contains some sulphur, and is therefore vulnerable to damage, but it is the effects on the nerves that are most worrying. Common symptoms of the effects of mercury on nerve cells included the shakes, irritability, paranoia and irrational behaviour. To add to the problems, the similarities between the mercury ion, Hg^{2+}, and Ca^{2+}, were close enough for mercury to be incorporated into bones. Mercury could stay locked up inside bones for the remainder of the individual's life, but the skeleton was a relatively safe place to store the metal out of harm's way. The real danger was caused when the body tried to get rid of the mercury *via* the kidneys. The damage it caused to kidneys was the usual cause of death because when they stop working, the buildup of urine causes the body to poison itself.

That syphilis patients would be associated with secretive behaviour and relegated to the margins of society is not surprising. Vampires could be a useful metaphor for the disease and its carriers, when polite Victorian society would rather not discuss such things openly. It has even been suggested that Bram Stoker had syphilis and was killed by it. The accusation, made by Daniel Farson, his great nephew and biographer, said the Stokers had a loveless marriage and Bram turned to prostitutes. Anyone in Stoker's position would certainly be discreet about visiting prostitutes, and physicians, out of a sense of propriety, were undoubtedly vague when it came to diagnosing syphilis. However, there is no evidence that he did visit prostitutes. The extremely weird final novel he wrote, *The Lair of the White Worm*, was taken as evidence of the mental effects of the disease, but it could just be a weird novel. His death certificate listed three causes: 'Locomotor Ataxy 6 months, Granular Contracted Kidney. Exhaustion.'

Exhaustion may not be considered a very scientific term, but it was probably accurate. Stoker's job at the Lyceum Theatre was very demanding, and he also found time to write novels. If he was then spending whatever scarce free time he had with prostitutes, exhaustion is hardly surprising. Locomotor ataxy is a loss of control over precise movements and can be a symptom of tabes dorsalis, a common condition that appears in the final stages of syphilis. It's suggestive, but not conclusive. Were the kidney problems a result of mercury treatments he had taken? Maybe. Maybe not. That Stoker wrote a novel that can be interpreted as a metaphor for syphilis could easily be coincidence, but it certainly shows how vampire fiction has gained particular popularity in times of epidemics, even in recent times.

The emergence of a new disease that could be transmitted through blood or sexual contact, and predominantly killed sections of society that were already ostracised, coincided with a surge in vampire novels, films and TV programmes. In the 1980s and 1990s, AIDS victims were demonised and feared. The Human Immuno Deficiency Virus (HIV) devastates the immune system to the point where the body is vulnerable to attack from bacteria, fungal infections and other pathogens. Before effective anti-virals became available, being diagnosed as HIV positive was a death sentence. That death could be brought about by infections that would be considered trivial in anyone with a fully functioning immune system. Full blown AIDS can leave people with visible scars and wasted bodies. Anyone living in the UK in the 1980s can tell of the nightmare-inducing public health adverts that featured gravestones and death knells that were broadcast to warn people of HIV and AIDS. It was a terrifying time.

There are some very uncomfortable parallels that could be made between the portrayal of vampires and common misconceptions about AIDS victims and how they were treated at the time. In the early years of the AIDS epidemic, as with many other epidemics before and since, misunderstanding and misinformation lead to stigmatisation and distrust. The responses and fears of villagers dealing with a strange new disease in the eighteenth century are not so far removed from modern responses to new emerging infections. Mass education programmes and intensive scientific research have gradually shifted perceptions of HIV and AIDS. Being HIV positive is now considered a chronic condition that can be effectively managed with medication. Thanks to high-profile campaigns and visible and vocal support from prominent individuals, attitudes to HIV-positive individuals have been transformed. They are not to be feared or shunned. The use of the sinister and malevolent aspects of vampires as a metaphor for syphilis or HIV is outdated, but other vampire characteristics still have something to tell us about the spread of sexually transmitted diseases.

8.4 MODELLING VAMPIRES

Disease-carrying microorganisms have the same impulse to survive and propagate their species as any other living thing.

Often, these disease-causing pathogens cannot survive or reproduce outside their host. Therefore, the host must be kept alive and able to spread the infection so that the microorganism can colonise new hosts.

Understanding how diseases spread is an important area of research that helps to contain outbreaks and minimise harm. Scientists build models that try to predict how different diseases are transmitted through societies using knowledge of how the pathogen enters the body and how long it incubates before symptoms appear, amongst many other factors. These models can then be used to see how epidemics can be best controlled and limited in their impact using vaccinations (if available), quarantines or other preventative measures to minimise the number of people affected.

Vampirism is a good model for how a sexually transmitted disease can spread through a population because a common way to create a vampire, in both folklore and films, is through bites. Biting someone requires intimate contact between two people, or one person and a vampire, making vampirism more difficult to contract than, say, air-borne or water-borne diseases. Even so, the rate of spread can be frighteningly quick. If a vampire bites just one person per day and that person becomes infected, the spread of the disease becomes exponential. On the first day there would be one vampire, and two vampires on the second day. By day three there would be four vampires and after only two weeks there would be over eight thousand vampires. If the spread continued unchecked in that way, it would take a little over a month for everyone on the planet to be transformed into a one of the blood-sucking undead.[||] But this is a very simplistic model.

One big assumption it makes is that, after contracting the vampire virus or bacteria, an individual is converted into a vampire and is able to pass on the infection within 24 hours. Many diseases have a longer incubation period and may not be contracted at the first exposure to the pathogen. Bacteria and viruses must pass between individuals in large numbers so that enough will survive the initial attack from the host's immune system to cause a viable infection.

[||] James Matheson's novel *I Am Legend*, showed the consequences of the unchecked spread of a vampire plague – Robert Neville is the only man left.

This simple model also assumes that vampires can get close enough to a victim every day in order to pass on the contagion. Zombies have a similar method of propagating their kind, but their shuffling, rotting bodies are easier to spot and run away from. Modern vampires have overcome this problem by being very attractive. Nevertheless, as the vampirism disease spreads, it will become more and more difficult to find new victims who aren't already infected.

Finding food would also become harder, as there would be fewer humans to feed on. Perhaps the growing hoards of vampires would have to resort to drinking animal blood (raising the question of whether you could have vampiric cows and sheep), or feed on each other. Such a rapid spread would quickly remove the pathogen's means of survival. There must be a balance between the numbers of predators and prey otherwise one, possibly both, will become extinct.

Again, modern vampires seem to have found a way round this by choosing which of their victims will become a vampire. According to films and fiction, it is not always enough just to be bitten, the victim also has to bite the vampire, or suffer repeated attacks from them. In other cases, a victim of a vampire attack will only transform after their death and can spend however many years or decades of their remaining natural life as a human. This would potentially enable a balance to be achieved between predator and prey.

Another factor to take into account is retaliation. Human immune systems fight off invading pathogens, and humans fight off vampires. The immune system tags invasive microorganisms with antigens to prevent them entering cells, and humans use garlic and crucifixes to ward off vampires (more on these preventative measures in Chapter 11). White blood cells engulf bacteria and viruses to destroy them, and humans use stakes to destroy vampires (see Chapter 12). But these defensive measures work both ways. Vampires, unlike zombies, are usually credited with high levels of intelligence that can help them evade the measures humans use against them. Many bacteria and viruses have also developed ways to evade immune systems. All of these factors slow the rate of transmission.

However, even with highly organised responses to the growing epidemic, humans would most likely be fighting a rearguard action, and it would only be a matter of time before a majority were transformed, leaving only isolated pockets of humanity. Modern vampires have the trump card in that, unless they are deliberately 'killed', they are immortal, and can presumably wait out periods of famine. This is similar in some respects to how some diseases have animal reservoirs and only occasionally jump to humans; for example, rabies in dogs and bats, influenza in birds and bubonic plague in rodents.

Incubation times, how easy or difficult it is to contract an illness, and many other factors have to be taken into account when building models to make them as accurate as possible. If vampirism was a disease that could be transmitted through bites then it has the potential to spread at a rapid pace, even taking into consideration a slower rate of transmission and all the preventative measures that could be taken. If vampires did not act to restrain themselves and preserve a large portion of the human population for food, then vampirism could easily become a global pandemic. But what if vampirism was more than a metaphor for the spread of disease. What if vampirism is the disease?

8.5 'AND AT THAT TIME BEQUEATH YOU MY DISEASES'**

Rituals around death and preparing bodies for burial are full of practices that protect the living from contracting an infection from the dead. They have also been interpreted as ways of preventing the dead from becoming vampires.

Dead bodies themselves are, of course, a potential source of infection. If the individual died of a contagious disease, there is a real risk of contracting that disease from it. Even if the person died as a result of an accident, there is still a potential risk of contamination from a body that is decomposing. We live with millions of bacteria in an ongoing battle between them and our immune system. After death, the immune system will no longer function and bacteria can multiply unimpeded.

Elaborate rituals have been established by humans that may, on the face of it, appear superstitious, but have potentially evolved to prevent infection from dead bodies. For example, in

**Pandarus, *Troilus and Cressida*.

some cultures where it is friends or family that traditionally prepare the body for burial, the water from washing a corpse must be carefully disposed of. In many cultures across the globe, the work of preparing the dead for burial or cremation is carried out by specialists in the community, who are experts in suitable methods. Professionalising care of the deceased means dead bodies are removed from everyday society, out of sight and out of reach of potentially harmful bacteria.

A new disease or a new outbreak could easily be transmitted to living members of a population through handling the dead. It is the kind of pattern that was seen in many of the famous vampire cases of the 1720s. A disease emerges in a village or town and several people die. With no knowledge of bacteria or infection, blame is laid on the first person who died and those who would have come into closest contact with the deceased are the next to be affected.

The 'vampirism as disease' metaphor is very common. Even the vampire-like *penanggalan* of Malaysia – a flying head with entrails streaming behind it that feeds on young children and pregnant women – may have very old and very long roots in explaining high infant mortality rates. The creature may be a way of reconciling the significant risks associated with pregnancy and child-birth, as well as post-partum infections, that were particularly common in the past. As we have seen, coincidences are easy to find between the symptoms of many diseases and the characteristics of vampirism.

Disease can dramatically change us both physically and mentally. Could it be that there is a vampirism disease, like that in *I Am Legend*, that changes its host so that they more closely resemble a vampire than a human? Viruses insert their DNA into their host's DNA in order to replicate themselves. The free and easy swapping of genetic information is part of what makes viruses hard to treat. The rapid genetic mutations HIV can undergo was one of the main factors that made finding effective treatments so difficult. Influenza mutates so quickly that new vaccines are needed every year. Could there be a virus that inserts vampire DNA into human chromosomes changing not just the virus but us too? The possibilities of the genetics of vampirism will be discussed in the next chapter.

CHAPTER 9

Species

> 'There were some good points about the – man, we were going to say – and yet we hardly feel justified in bestowing upon him that title'.
>
> James Malcolm Rymer, *Varney, the Vampyre*

Miriam Blaylock's needs were very simple: six hours sleep in every twenty-four and the blood of one human being every week. With this regular routine, Blaylock had survived for thousands of years. To all outward appearances, she was an ordinary human being, but her physical form disguised the fact that she was something else entirely.

Whitley Strieber's 1981 novel, *The Hunger,* tells Blaylock's tale – how she had evaded those who wanted to hunt her down, lived through periods of persecution and did everything possible to pass as human, but she was the last of her species. Unable to die and unable to reproduce, she went for the next best thing, transforming carefully chosen human companions into vampires like herself. But her transformations were not perfect; unlike her own immortality, her companions' health deteriorated after a few centuries. Looking for explanations, she stumbled across research scientists studying gerontology (the science of ageing) and submitted herself for tests. The results revealed just how different she was: 'The gross abnormalities of the blood and the completely alien brain function left no room

Vampirology: The Science of Horror's Most Famous Fiend
By Kathryn Harkup

Published by the Royal Society of Chemistry, www.rsc.org

for doubt: Miriam Blaylock was not a member of the species *Homo sapiens.*'

The DNA extracted from Blaylock's blood, and her physical resemblance to humans suggested that, though she likely shared a common ancestor with humans, Blaylock's evolutionary path had been very different. Vampires as a similar but distinct species from humans is a common theme in modern vampire literature and perhaps it is a nod to the vampire-like creatures of ancient times that could appear human-like but were actually something else (see Chapter 1). In *The Hunger*, Blaylock claims her mother was Lamia, the shapeshifting femme fatal of ancient Greece, but her earliest memories were from ancient Egypt.

The eponymous antihero in *The Vampire Lestat* also traced his heritage back to ancient Egypt, claiming a vampire lineage originating with Isis and Osiris, or at least creatures that had adopted those names. These beings were happy to use the myths that surrounded Isis and Osiris as cover for their vampiric nature. Lestat learns that they obtained their vampire qualities when their bodies were combined with that of a demon, perhaps indicating they are more like a hybrid than a different species. This demon spirit could also be passed on to other humans through a vampire's blood – suggesting inheritance through blood-lines.

The original ancient Egyptian myths associated with Isis and Osiris are complex and varied, but one of the most common versions has King Osiris killed and cut into pieces by a jealous brother. Isis, Osiris's wife, collected up the pieces and reanimated him using magic, and conceived with him a child, Horus, who inherited the throne from his father. Osiris, no doubt because of his practical experience, became king of the underworld, where he presided over the dead. But apart from returning from the dead, there is nothing to link any of these well-known Egyptian figures with vampires. The mysteries, glamour and human/animal/god hybrids of ancient Egypt seem to hold a strong draw for vampire fiction writers despite there being no evidence for vampire-like creatures in their traditions.

Nevertheless, this doesn't detract from the fact that vampires are often considered not to be human. What if vampires appear and behave differently to humans because they really are different? There are a few species classed as hominids alongside humans, such as chimpanzees, gorillas and orangutans, but humans have long held themselves distinct from the rest of the animal kingdom. This may be because we look and behave differently to our fellow hominids, but there is probably more than a little ego contributing to our sense of superiority. Humans are undoubtedly successful; they dominate over other hominids, and not just in terms of numbers. Could it be possible for another human-like species to exist alongside us?

9.1 EVOLUTION

There are still many mysteries about how life began on this planet, but it evidently happened somehow, because the result is plain to see all around us. What is very clear is that there is no single way for life to survive on earth. To ensure that the show goes on, life has adapted and diversified to survive in almost every conceivable environmental niche that can be found on this big ball of rock spinning around the sun. Is it such a stretch to think there might be a species that looks almost exactly like us but behaves rather differently, and has managed to evade scientific scrutiny?

Finding species previously unknown to science happens all the time. Take the example of beetles: the writer A. A. Gill once joked that in the Amazon rain forest it was only beetles found on trees between human waist and head height that had been properly studied. There were so many new beetles to be found and identified, that scientists were kept busy enough without having to go to the effort of bending down or straining their neck to find more. And that's just one type of animal in one part of the world. The fact is that humans have barely scratched the surface when it comes to identifying all the varieties of life we share our planet with. Scientists aren't even sure how many different species there might be yet to find.

Humans or humanoid species are obviously bigger than beetles and might therefore be expected to be easier to find. But, as was shown in Chapter 7, appearances can be deceptive, and some creatures can hide in plain sight by disguising themselves as other animals. Just because one animal looks a lot like another, it doesn't mean they are closely related.

Two different species can be superficially similar, but genetically very different. For example, elephant shrews, small furry animals with long noses, have more in common with elephants than shrews, genetically speaking. Animals have to adapt to their environment, and a particular appearance might help an animal survive in a particular environment. Two different species can arrive at the same point or appearance because it suits their environmental niche, but by following very different paths. It might therefore be possible that two humanoid species could look very similar but also be very different, rather like humans and vampires.

The idea of adaptation and slow changes in successive generations of living things has been around for a couple of centuries. In 1794, Erasmus Darwin wrote *Zoonomia*, where he speculated 'that all warm-blooded animals have arisen from one living filament.' Sixty years later, his grandson Charles penned *On the Origin of Species,* in which theories of evolution were expanded and consolidated in the phrase 'natural selection': 'one general law, leading to the advancement of all organic beings, namely, multiply, vary, let the strongest live and the weakest die.' But it wasn't until 1952 that Erasmus Darwin's 'living filament', the thing that passed on these changes to future generations, was identified as DNA by Alfred Hershey and Martha Chase. It is cumulative changes to the DNA molecule over successive generations that result in the divergence of living things into different species, and occasional convergence of their physical characteristics that can make them appear very similar.

9.2 DNA

Deoxyribonucleic acid (DNA) is perhaps the most important molecule on earth. The same molecule is used as the instruction manual for being by every known living thing on the planet. For all the complexity DNA can generate, it is relatively simple in its construction. Four units, known as base pairs – Adenine (A), Thymine (T), Guanine (G) and Cytosine (C) – are strung along a backbone made of sugars and phosphates. Each strand of DNA is complimented by a second strand and the two are linked together by hydrogen bonds, a little like a zip closing. The teeth of a zip, and the base pairs of DNA, have to match each other in terms of shape. Base pairs also match in terms of the placement of particular atoms meaning base A complements base T, and G matches C. Therefore, if G is found on one strand, there will be a corresponding C on the opposite strand. The complementing of bases allows a neat alignment of hydrogen bonds to form and hold the two strands together.

As a result of these elegantly matched base pairs, DNA forms long, smooth, double strands that curl and twist into compact forms so that they can squeeze into the tiny nucleus of cells.[†] The DNA held within each cell is the complete set of instructions, like a library of information, that can be consulted for the construction of a living thing.

Instead of one continuous length, DNA is broken up into smaller units called chromosomes. Laid out neatly in rows, chromosomes look a little like pairs of stripy socks (see Figure 9.1). Humans have 46 chromosomes, or 22 pairs plus two sex chromosomes (XX if you are female and XY if you are male).[‡] Different animals can have different numbers of chromosomes, for example, chimpanzees have 48 and goats have 60. In *The Hunger*, Miriam Blaylock's DNA was found to be split into 68 exceptionally long chromosomes, clearly demonstrating that she was not human. The number of chromosomes is important, particularly when it comes to producing offspring, but it is the information contained within the chromosomes that shapes our characteristics, from eye colour, to the ability to digest different foods.

Within the chromosomes are genes, a set of instructions to build specific proteins or enzymes.[§] To expand on the sock analogy, the genes are the individual stripes on the sock, or, in the library analogy, this would be a specific book on a shelf. In order to read the instructions contained within the gene, the two DNA strands must be gently prized apart to access the sequence of base pairs. There are sequences of bases that signify the beginning of a set of instructions, which direction they should be

---

[†]There are a huge number of single celled organisms that do not have a cell nucleus, prokaryotes such as bacteria, and their DNA floats free. The members of this branch of the tree of life carry their DNA in little loops that float around inside the cell.

[‡]There are exceptions to these general rules, for example, Down's Syndrome is caused by the presence of an extra chromosome, and some people have extra sex chromosomes.

[§]There are several definitions of what a gene is or does, but we will stick with this one.

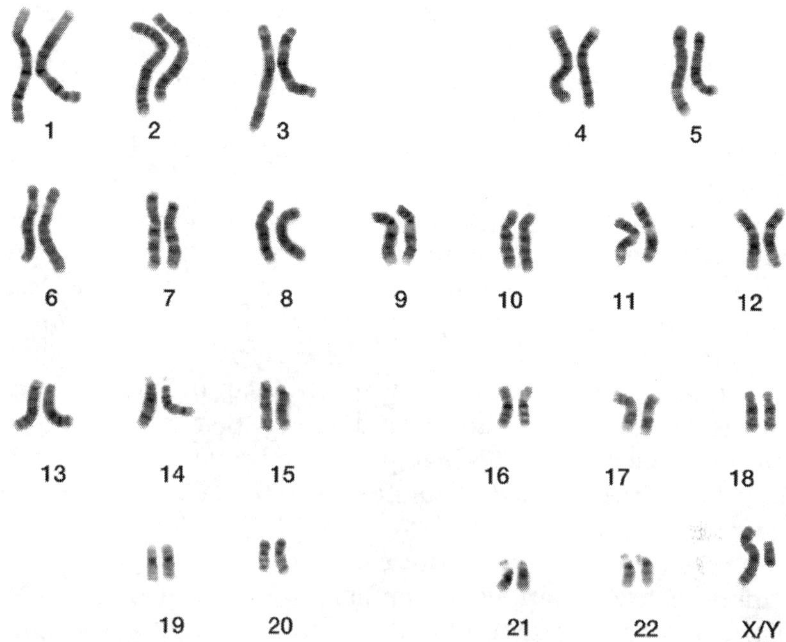

Figure 9.1 Showing a set of human chromosomes from a man arranged into their pairs.
Credit: National Human Genome Research Institute, https://www.genome.gov/genetics-glossary/Genome.

read in and where to stop reading. The instructions themselves are very straightforward.

Combinations of three bases signify different amino acids: GCT means alanine, GGT means glycine, *etc.* Reading along a strand of DNA is a simple step-by-step list of which amino acid to add next to the growing chain that will eventually become a protein or enzyme. Molecular machinery within each cell reads the code and constructs the strings of amino acids accordingly. This is, of course, a very simplified overview of what is actually going on inside cells, but hopefully serves our purposes going forward.

Every time a cell divides, a complete set of DNA instructions has to be copied for the new cell. Obviously, errors can occur in this process, sections can be missed out, swapped round or copied twice, but there are all sorts of checks and balances to keep these errors to a minimum. There can also be chemical damage to the DNA strand or damage from ionising radiation like ultraviolet light, and again there are mechanisms to minimise the damage and repair things where possible. These changes are called mutations and they occur all the time. For the most part, these mutations have little or no impact, good or bad; the cell can patch up the damage so that life can carry on as before.

A simple mutation can cause a change in the sequence of base pairs in a gene and, therefore, result in a change in the amino acid sequence of an enzyme. This might make no difference to the enzyme, but it might stop it from performing its function, or it could enhance its ability to do whatever it does. If that slight change confers some kind of advantage to an individual, it can increase the chances of survival and passing that advantage on to a future generation. But that genetic advantage will only be passed on if the beneficial mutation is in the DNA within sperm or eggs.

Over staggeringly long periods of time and many, many generations, these mutations accumulate. As the sequence of DNA instructions diverges, so too do the animals that carry them, eventually becoming distinct species. The more differences there are between two samples of DNA, the more distinct are the two creatures they were taken from. A few changes here and there and maybe one person has slightly sharper teeth – all the better to bite you with, my dear. A slight slip of the sequence over there, and maybe there are alterations to the proteins that act as melatonin receptors. This could change how an individual reacts to the daily build-up of melatonin in the body, making them sleepy during the day and wakeful at night.

Genetic differences have obviously contributed to the wide range of differences that humans display, from skin colour, to hairiness, from eye colour to smelliness, but the differences are superficial. At the genetic level, the variation is so slight that all humans living on the planet today belong to the same species, regardless of outward appearances. That hasn't necessarily always been so.

In the distant past, there were different species of humans walking around and interacting with each other. If it was true then, could it be true now? Could it be that *Homo sapiens* are not the only *Homo* species on the planet and we coexist with members of *Homo vampiris*?

9.3 HUMAN EVOLUTION

Despite the fact that Charles Darwin wrote *The Descent of Man* in 1871, it is only relatively recently that theories have been established on the path of human evolution and there is still a lot of academic debate on the subject. The relationships between the different branches of the *Homo* evolutionary tree are not always clear and much is speculative.

In general terms, around 700 000 years ago, a species or subspecies of archaic humans emerged, now known as *Homo heidelbergensis*. Later, *heidelbergensis* is believed to have diverged into *Homo neanderthalensis* (Neanderthals) and *Homo sapiens* (modern humans, or us),[¶] but the details of the exact evolutionary route taken are not clear. Both Neanderthals and *H. sapiens* coexisted up until the last of the Neanderthals died out sometime around 30 000 years ago on Gibraltar. Whether Neanderthals and modern humans were distinct species is just one of the many areas of debate around human evolution.

One problem is the difficulty of defining exactly what a species is. Experts juggle several different definitions, according to their area of interest and what works best in a particular situation. One of the most commonly cited definitions, and the one you are most likely to have come across in school, is if two animals can breed to produce fertile offspring they are considered to be of the same species, as is the case with modern humans.

However, there are problems with this definition. For example, bacteria don't reproduce sexually, but often freely exchange DNA; hybrid species also allow the interbreeding and extensive mixing of DNA that makes it difficult to form clear boundaries between closely related individuals, and the remains of extinct animals may show they were near identical in appearance, but this gives no indication of whether they could successfully breed together. How successfully modern humans and Neanderthals may have interbred in the past is not easy to establish. Traces of Neanderthal DNA have been found in humans living today, suggesting a possible remnant from interbreeding thousands of years ago. For some, this suggests that Neanderthals and modern humans are better described as subspecies, rather than truly separate species.

[¶]*Homo vampiris* excepted.

Trying to answer some of the questions around species brings up another problem: the paucity of skeletal remains of our human ancestors. Scientific techniques developed in recent decades have transformed the way archaeologists can obtain information about the finds that have been made. Carbon-14 dating and analysis of other trace elements that undergo radioactive decay, have vastly improved the accuracy of dating samples that scientists have available to them. DNA analysis can also help establish relationships between different groups of ancient humans. The rate of progress in understanding human evolution has accelerated in recent years but there is still much to learn. New discoveries are being made all the time, but this doesn't necessarily help clear up the confusion.

In 2010, human bone fragments were discovered in a cave in the Altai mountains in Siberia. The discovery caused a media sensation for many reasons. The individual that the bones belonged to was clearly human, but very different to any other human species previously discovered. This previously unknown human species was named Denisovan or Denisova, after the Denisova cave in which they were found. The few fragments of bone that have been excavated are not enough to give a reliable idea of what Denisova looked like; a finger bone that was found is of a similar size to that of a modern woman, but that is about the biggest clue available, and not much to go on. The DNA that has been extracted from the fragments, however, are more revealing.

Analysis shows the Denisovans shared a closer relationship to Neanderthals than to modern humans. At least three species, or subspecies, of human were therefore coexisting on Earth tens of thousands of years ago. Another discovery, of the remains of *Homo floresiensis*, nicknamed 'the hobbit' for her tiny one-metre-high stature, show that there are potentially even more ancient humans to be found. It shows humans could, and potentially still can, be very different to each other and coexist.

The variety of humans that have existed through Earth's recent history also shows how, like every other animal, they have adapted to their environment. *H. floresiensis*'s small stature may be because she was a descendent of some of the earliest African *Homo* species. Members of this African species perhaps left the continent and found a niche in Indonesia where they could survive long after their African forebears had become extinct. Another theory suggests that isolation on an island created environmental pressures that favoured a smaller stature. DNA analysis could potentially answer many of these questions but attempts at extracting DNA from the remains of *H. floresiensis* have so far been unsuccessful.

Initial studies of the cave in Indonesia where the hobbit was found indicated that *H. floresiensis* only became extinct 12 000 years ago. That is a blink of an evolutionary eye. The local population also knew old stories of small people who lived in the hills, and some wondered if *floresiensis* and her kin had survived to the present day, albeit only in traditional oral stories. Although the date of extinction was later revised to around 50 000 years ago, it is still astonishingly recent and certainly contemporary with modern humans. Given the variety of humans that have existed in the past, and the fact that some of them were completely unknown until the last few decades, how likely is it that there could be other species still to be discovered?

The discoveries of previously unknown hominid species certainly came as a surprise. However, these species or sub-species are extinct. Their remains have been hidden in the ground and there is undoubtedly more to discover. Vampires, however, though they spend a lot of time in their coffins, have to come out to feed at least once in a while, making them theoretically easier to spot. In *Dracula*, the Count's true nature was certainly known to the local population, but poor communication between rural Romania and the rest of the world meant he could keep a low profile for most of his centuries-long existence. Today, with international travel, global communication and non-stop news coverage, how difficult would it be for a vampire like Dracula to remain in obscurity?

9.4 A NEEDLE IN A HAYSTACK

There are several factors that could make vampires difficult to spot, such as how many individuals there might be, and where they spend most of their time. Any population needs access to food, protection from dying of disease or predatory attack, and a way of reproducing itself. Spotting a small number of vampires amongst billions of humans is obviously more difficult than if there were lots of them. Humans have survived surprisingly narrow population bottlenecks. What might be a viable population of *Homo vampiris*?

Looking first at humans, evidence of historical population sizes has been found in our mitochondrial DNA. The DNA found in the mitochondria structures within our cells is slightly different to the DNA in the nucleus because it is inherited from the maternal line only. Analysis of mitochondrial DNA shows that modern humans are descended from fewer than 5000 female ancestors. The total population at that time may have been much larger, as there may have been more men than women and not all of the women would have been breeding. Even so, if there were as many as 20 000 ancestral humans, this is a tiny number compared to the approximately seven billion people that currently occupy the planet.[||] With their excellent health and grossly extended lifespans, a vampire population could be much, much smaller.

Vampires, because they feed on humans, would need to be close to large populations of humans to sustain themselves, so we shouldn't have to travel far to find them. However, their nocturnal behaviour means they aren't necessarily up and about at the same time. There are somewhere around 4500 cities in the world with populations over 150 000. Even if there were 20 000 vampires, that would be only four vampires per city. A larger city could perhaps sustain a bigger vampire population, perhaps as in the TV series *Angel,* where Los Angeles is shown to have a thriving vampire and demon population. A tiny population of the blood-sucking undead spread thinly throughout pockets of dense human habitation might be viable, as well as difficult to detect. But could such a population have evolved separately from humans in the first place?

As a fun thought experiment, let's speculate as to how *Homo vampiris* could have split off from the *Homo* family tree and explore its chances of survival to the present day. How likely is the divergence of two human species tens of thousands of years ago, as Strieber depicts Miriam Blaylock's evolutionary backstory in *The Hunger*?

[||] For comparison, chimpanzee numbers are thought to be between 170 000 and 300 000 in the wild, and they are listed as endangered.

9.5 THE DESCENT OF VAMPIRE

Evolution does not progress at a regular pace. Changes in circumstance can drive rapid evolutionary adaptations. Climate change, predator numbers, food scarcity and disease can all put pressure on a species population and increase the chances of some genetic mutations surviving into future generations in higher proportions than others. Many such events and subtle changes have created the enormous diversity of life on planet Earth. What may seem unusual-looking animals to us are simply the consequence of what many generations of life has thrown at their DNA. Given how different life can look and how many different ways life has found to survive and thrive on this planet, is it such a stretch of the imagination to consider a form of life that looks human and feeds on human blood?

At some point in the very distant past, perhaps one group of early humans decided a rival group of early humans tasted better than mammoth, or were easier to catch than sabre-toothed cats. But it's a giant leap, or many, many small steps, from occasional cannibalistic snacks, to a grossly extended life living in the shadows and feeding exclusively on blood. Sleep patterns, digestive processes and lifespan would all have to undergo dramatic transformations. The switch wouldn't be made overnight, as depicted in many vampire films and TV programmes. However, minute adaptations building up over thousands of years is a more feasible, if slower, story.

Having acquired a taste for their neighbours, some small changes might be advantageous to their predatory nature. Being awake and alert at night might give them an advantage when preying on their rival humans. Subtle changes to the gut could make blood easier to digest. A slight increase in lifespan, from a better immune response or resistance to the ageing processes, would mean they could outlive their food and survive in smaller numbers. As we have seen in previous chapters, some animals have adaptations that would suit a vampire lifestyle, and these animals, however different they appear to be now, have all evolved from the same genetic pool. In many cases, the genetic information that gives these animals their vampire characteristics can be found within human DNA; it's just used differently.

First of all, let's look at the most defining vampire characteristic: food. Several species have genetic clues to how it might be possible to adapt to a blood diet. The vampire bats we met in Chapter 4, a fellow mammalian species and therefore fairly closely related to humans, has successfully adapted to an all-blood diet. The vampire squid from Chapter 7, has one of the slowest metabolisms yet discovered – a handy trait to have when only getting the occasional blood meal. But when it comes to infrequent feeding, pythons are the champions. Rather than gorging themselves ahead of a hibernation period as other animals do, they allow their organs to shrivel between meals to conserve energy, and rapidly expand them again when they feed. When pythons do eat, they ramp up their metabolism very quickly to digest their large meal before it rots in their stomach. A python heart and liver can be 50% bigger the day after a meal and the small intestine can double in mass in the same time. Two days later and the snake's organs start to shrink back to starvation mode.

This ability to rapidly change metabolic processes in the body means snakes can survive long periods without eating and not get indigestion when they do eat. The same ability would certainly help the bloodsucking undead when it comes to digesting huge gluts of blood. The kind of atrophy described in the snake organs during periods of famine are remarkably similar to descriptions in *The Hunger*. The vampiric creatures in Streiber's novel don't die if they are starved of blood, but shrivel up and waste away to husks of their former selves.

Experiments have shown that taking plasma from a recently fed snake and adding it to rat cells in a dish causes a sudden burst of growth. The rats' genes understand the same instructions to grow, even though they have come from a completely different species. The basic mechanism is there in rats, all that is missing is the means to switch it on and off. Studies have shown that this on/off switch in pythons is anything but simple, though the experiment with rat cells shows the system, whatever it may be, need not be exclusive to snakes or even reptiles.

Secondly, the secrets of longevity might also be found within our DNA. There is understandably a lot of interest in finding out how we age and if anything can be done to slow or stop it. Very generally speaking, the bigger the animal, the longer it lives. Elephants can live into their eighties, while most mice struggle to make it to their fourth birthday. There are exceptions to this trend, and by studying these unusually long-lived animals, scientists hope to find out what makes some species more resilient to the ravages of time. There are nineteen species that are longer-lived than humans, relative to their size. One of them is the naked mole rat, which we met in Chapter 7, and the other eighteen are all, coincidentally, very strongly associated with vampires; they are bats.

The *Myotis* genus of bats are the current record holders for longevity.** Of the *Myotis* species, the winner is Brandt's bat (*Myotis brandtii*), a small European bat that has been known to live for at least 41 years. This is roughly ten times longer than similar-sized animals, and, after scaling, approximately twice that of humans. The secret is thought to be the telomeres in the bat's DNA.

Telomeres are sections of DNA that act as caps at the end of chromosomes to stop them unravelling. Every time a cell divides, the DNA is duplicated and a little bit of the telomere gets lost. When the telomeres get too short, the cell tends to shut down. This effectively sets a limit on how often a particular cell can divide. Normally, the age of an animal can be fairly accurately predicted using telomere length, but not when it comes to bats; their telomeres never seem to get any shorter.

One way of maintaining telomere length is to make more and add it on to the end. There is an enzyme, telomerase, that can extend the length of the telomeres, increasing the number of divisions and therefore the lifespan, or replication number, of the cell. Experiments in mice show that animals that don't have functioning telomerase age quickly. The code for this enzyme is in human DNA in every cell nucleus in our body, but not all cells use it. Hair follicles, bone marrow cells and skin cells, for example, need telomerase because they divide frequently. Other cells don't necessarily need telomerase because they don't divide as often. And, as we age, the activity of the enzyme decreases. It would therefore seem obvious that the way to slow ageing would be to increase telomerase activity in all cells and we could all look forward to the longevity enjoyed by Count Dracula and his fellow vampires. Scientists have the ability to

**There are many genuses of bats still to be investigated.

increase enzyme activity, but they have learned that it comes with consequences.

One of the ways we know telomerase works effectively in making cells divide indefinitely is because it is found in cancer cells. It is the telomerase that contributes to the unregulated growth of these cells. Bats rarely get cancer and studies have shown that their telomerase activity is low, as in healthy human cells. The bats have another, as yet unknown, way of keeping their telomeres the right length and their DNA healthy throughout their long lives.

Telomere length is not the only thing that determines how many times a cell can divide. How cells age is evidently a complex process, but several clues have been unearthed. One is the enzyme Sir2, that comes in seven different mammalian varieties – SIRT1 to SIRT7. Yeast, certain types of worm and fruit flies that over-express this enzyme live longer.

Sir2 acts on histones, structures that help DNA fold into its compact shape so that it can all fit inside the nucleus of the cell. Bits of DNA that aren't needed as often get wrapped up tighter around histones and other bits are left more loosely wrapped so they can be accessed and read more easily. Acetyl groups help control how tightly the histone wraps the DNA. Therefore, the more acetyl groups on a histone, the looser the DNA is wrapped. SIRT6 seems to target histones near telomeres, removing acetyl groups and thereby making the telomere bind more tightly to the histone. This could be how the telomere is protected from shortening when the cell divides.

There are many more complex and interrelated mechanisms at work inside cells that keep them healthy and dividing for longer. There are also external factors that have to be taken into account that can potentially undo all the good work done by Sir2 and telomerase. Stress may play a significant role in how quickly telomeres get shortened. Reducing stress may be a better way to slow ageing than manipulating enzyme activity in cells. A vampire's prolonged undead existence may be due to the extensive amount of rest and relaxation they enjoy, slumbering in their coffins between meals.

Another factor is diet. Animals living on restricted calories live longer. Perhaps restricting themselves to a low-calorie, blood-only diet is another way vampires have found to extend their existence. However, cutting down on cake and biscuits is not a short cut to eternal life. In yeast experiments, restricting access to glucose from 2% to 0.5% extended their lifespan by 30%. Rodents fed on 60% of their normal calories live longer, but not forever, and the restrictions have to be in place throughout their extended life to see the benefits.

As we understand more and more about what different sections of our DNA do and the relationships between competing systems, it soon becomes clear there is no easy route to eternal life. The systems that preserve our DNA and the telomeres at the ends of chromosomes have to be kept in check. A magic pill containing telomerase to replenish our stocks might increase the length of our life, but at the risk of shortening it in the form of unhindered cell division, or cancer. Ageing seems to be a trade off between a long life and a healthy one. Bats and naked mole rats may have found a recipe within their genes that allows them greatly extended life, and almost completely cancer-free, but even they are not immortal.

While the mechanism that gives bats such a long life-span is a mystery, there are many other genetic traits that are well understood. If these genes are also present in human DNA, why aren't we more like bats, or vampires? The accumulation of subtle mutations over millions of years of evolution means there are differences between genes, but this does not account for all the differences between species. It is not just what the genes code for that is important, but how those genes are expressed. DNA instructions are open to a certain amount of interpretation. If humans already have the genetic instructions that could result in vampire-like characteristics, perhaps the trick is learning how to switch them on.

9.6 READING THE INSTRUCTIONS

Having a gene that codes for a particular enzyme does not automatically mean that enzyme will ever be made. Different genes are needed in different situations and at different times, and so there have to be ways to switch genes on and off as required. Sometimes a gene needs to be permanently switched off, for example, during foetal development in the womb. All of us have grown from a single fertilised egg into our complicated forms. As a foetus develops from a single fertilised cell to billions of cells, those cells have to specialise. Heart cells have a very specific function: they must contract to create a heartbeat. Cells in our retina at the back of our eye must be sensitive to light, and absolutely must not contract or everything would appear very blurry. Yet both the heart cell and the cell in the retina contain the same DNA information. Some of the information in our chromosomes has to be ignored and other bits amplified to make sure the cell in question does the right thing. And, once a cell has decided to become a heart cell, there must be no going back, we don't want eyeballs suddenly growing in our hearts.

Although it might make sense to edit out the bits of DNA that aren't needed as cells start to specialise, this is risky. A simple slip of the molecular scissors could remove something important and the cell could die. And, just because a cell doesn't need that information at that precise moment, doesn't mean it never will. A heart cell might need to speed up or slow down its contractions on occasion, but it would be hoped that it could be relied upon not to change its behaviour too much, as that would have dire consequences. Other cells have to go through more drastic changes. Our skin cells need to increase the production of melanin when they are exposed to sunlight. Cells in our pancreas need to produce insulin in response to the amount of sugar we have eaten and not produce a constant amount all the time.

If a cell's DNA is a library of all the information available to it, each cell must know how to read the right bits and how often. This is the realm of epigenetics, meaning 'on top of genetics', the modifications made to our genetic material that change the way genes are switched on or off without changing the genetic code itself. This branch of genetics is still in its infancy, but scientists have discovered a staggering amount in a short space of time. The word was first coined in 1942 by Conrad H. Waddington to describe the influence of genetics seen during development and steady progress was made during the twentieth century, but things really took off in the 1990s.

Along with coding DNA – the bits that have instructions for making proteins and enzymes – there is also a lot of non-coding DNA. For a long time this was thought to be junk, strings of genetic nonsense that the cells ignored but for some reason didn't prune out. It is now known that a lot of this non-coding

DNA does have a function, and that is to regulate the coding DNA. We are only just learning how the function of different genes is controlled and organised.

One of the ways genes can be switched off, or their expression reduced, is using the histones we met earlier. By wrapping up sections of DNA, it becomes more difficult to access it, effectively switching off the genes found in that part of the sequence. Another way is through methylation. In theory, adding methyl groups to regions of DNA switches off genes, and it is true that unmethylated regions of DNA are a hive of activity, but things are not so clear-cut. There are many other factors at work to regulate how our genetic code is utilized, but what is certain is that the code can be chemically annotated or modified by environmental factors. Those factors might be whether a cell is near the surface or interior of a growing foetus, or whether a cell is exposed to sunlight or a particular hormone. Even the food we eat can make a huge difference to which genes get read more often than others. A vampire hidden away from sunlight and feeding exclusively on blood exists in a very different environment to most humans. It might not be such a stretch to think this would influence how their genes are expressed and give rise to some of their distinctly non-human characteristics.

As an example of how powerful environmental factors can be, take bees. In bee colonies, there are thousands of bees that are genetically identical, but some of them look very different. In the early stages of development, nurse bees feed bee larvae with royal jelly. At around three days old, the bee larvae are switched to a diet of pollen and nectar, and these will grow up to be worker bees. But a few bee larvae, though they may be genetically identical to their fellows, are selected to continue their diet of royal jelly. These bees will grow up to become queen bees, and will continue to feed on royal jelly throughout their royal lives. The changes that occur must be epigenetic and the changes brought about by food. Could an all-blood diet be the determining factor in switching on key genes that give vampires their characteristic traits?[††]

[††]Even if it could, the nutritional deficiencies would quickly counteract any epigenetic benefits. In other words, it would cause serious health problems before any benefits might appear. Best not to try it.

Switching on redundant genes, increasing the expression of some and dialling down the influence of others, could potentially result in a human showing more vampire-like characteristics. But what if the necessary genes simply aren't included in the DNA sequence. Could genes be borrowed from other species? Was Anne Rice on to something in *The Vampire Lestat* when she described the origins of vampires as a hybrid between a demon and two humans? Could the demon simply be a metaphor for genetic information borrowed from another species and incorporated into human DNA? In other words, could humans become vampires quickly by genetic modification rather than taking tens of thousands of years to evolve? The answer is maybe. It's complicated.

9.7 GENETIC MANIPULATION

Genetic modification is the science of altering specific bits of DNA, such as removing, adding, amplifying or modifying certain genes. Because all living things on Earth use the same genetic language, it is theoretically, and practically, possible to swap bits of DNA between totally unrelated species. This is known as transgenics, and there have been some very notable successes. For example, diabetics rely on injections of insulin to regulate their blood sugar levels because their own body can't produce insulin. Insulin is a protein that would be difficult to produce in a lab and far too time-consuming and costly to meet the demand. Instead, the gene that codes for insulin can be isolated from human DNA and inserted into a bacterium so it can do the work for us.[‡‡]

Bacteria and humans are at very different points of the tree of life, but DNA is DNA and the bacteria simply follow the instructions and start producing human insulin. Bacteria reproduce at a significantly faster rate than humans and so the production of insulin can be quickly scaled up to meet the needs of diabetics. If this can be done with bacteria, why not simply deliver the required gene directly into the human diabetic and they could start making their own insulin? This method of

[‡‡]Not all insulin is produced this way – other insulins are available.

treatment is sometimes called gene therapy and, unfortunately, it is not so simple.

To deliver the DNA into a bacterium requires a vector and the chosen vector in genetic modification is often a virus. Viruses are perfect for the job, because they are masters at finding their way inside cells and inserting their DNA into the host's DNA. In the case of insulin, the virus and the bacteria never go near the diabetic patient, as it is only the insulin that is needed. To insert the required gene into the bacteria, a virus is chosen that only infects bacteria cells and not animal cells, further reducing the risks. While this solves one important potential health problem, it creates other difficulties.

Though our genetics can cause us a lot of health problems, there is rarely just one gene at fault. Even when a single gene can be identified as the defining problem, as in the production of insulin, and if the virus chosen as the vector for humans is low-risk health-wise, there is no guarantee of success. Cells are wise to viruses' tricks. In our body, immune systems work hard to prevent viruses infecting cells in the first place. Some viruses can clearly evade these defensive tactics and so more checks are carried out inside cells to look for foreign DNA and eliminate it when cells divide. All of these factors have to be overcome if the insulin gene is to be delivered directly into a diabetic patient. Viruses are also known for their ability to mutate rapidly and many researchers are exploring alternative ways of delivering DNA to avoid infections.

On top of this, the gene also has to be delivered to the right cells. It is no good having cells in the mouth or kidneys producing insulin, it would be in the wrong place. Even when a single gene has been identified, and using a delivery system that is harmless to humans, and injecting those viruses directly to the site they are needed, it does not necessarily mean the required gene will make it into the right type of cell, be inserted into the human DNA, read and used to make insulin. Using the diabetic example, if all went well and cells started to produce insulin, the effects could still wear off. During the normal cycle of a cell's life, DNA is checked, errors are corrected and foreign material removed. Insulin production can grind to a halt at a later stage. This is why gene therapies are not necessarily the magic bullet they sometimes appear to be. There is also a very real risk of cancer, as has been found to be a result of some early gene therapy treatments. Nevertheless, advances in genetic modification techniques, and gene therapy, means more of these problems are being overcome to offer safe and successful long-term benefits for human health. But it is very early days.

Viruses as vectors are a slightly old-school technique in terms of methods of genetic modification, but it might explain how vampires pass on their vampiric qualities to humans when they bite them. If vampirism itself is the disease, as discussed in the previous chapter, then a vampire virus might be capable of inserting genes or modifying DNA in its new host to express vampiric characteristics.

A more recent development, CRISPR (clustered regular interspaced short palindromic repeats), allows a more direct intervention into the genetic code and avoids many of the troublesome aspects of using a virus as a vector. CRISPR are DNA fragments from organisms that don't have a cell nucleus, such as bacteria, that are used to delete foreign material from their DNA as part of their defence system. When combined with the enzymes Cas9, that recognise specific sequences of DNA, CRISPR-Cas9 techniques can be used to edit genes in individual cells. It has huge potential, from boosting crop yields to modifying mosquitoes to prevent the spread of malaria.

There are also benefits on the human scale. A few cells can be taken from an individual with a genetic disorder and the mutated sequences edited. These modified cells can then be grown in a lab and reintroduced into the host. Initial trials are promising, but there is still a long way to go. However, it is theoretically possible to use CRISPR to create a vampire. The first step would be harvesting a few human cells, perhaps skin cells that are easy to work with in the lab, and editing the DNA to introduce vampire characteristics. These skin cells could then be multiplied and reintroduced into the donor human. Alternatively, these same skin cells could be coaxed into becoming sperm or egg cells, which could then be fertilised and implanted into a uterus to make your own IVF vampire.

When Streiber wrote *The Hunger*, in 1981, gene therapy was just a theory, no clinical trials had yet been carried out, and wouldn't be for more than ten years. What he described when his vampire character Miriam Blaylock transformed one of her human partners was a blood transfusion. Bone marrow transplants and solid organ transplants do introduce the donor's DNA into the recipient's body. Because donor and recipient are carefully matched for compatibility, and immunosuppressors are available, the body can tolerate the new organs and the new DNA.

In the case of bone marrow transplants, normally the recipient's bone marrow is first destroyed, and their immune response along with it, before donor cells are grafted in. This minimises the risk of rejection, though immunosuppressors are still needed, and the new bone marrow takes over. This means the recipient gets a whole new blood system: red blood cells, white blood cells, platelets and all.

In *The Hunger*, the cells within Blaylock's blood had the ability to consume human blood cells and reproduce more of themselves without alerting the host's immune system to their presence. There is no mention of bone marrow or tissue matching her partners, but the end result is the same, a completely different type of blood operating within the human system. Her once-human partners became human-vampire chimeras.

The prospect of tinkering with human DNA to produce vampire-like characteristics through evolution, epigenetics, viral infection, gene therapy or transplants, is certainly intriguing. If we are being generous, novels like *The Hunger*, or *I Am Legend* where an infection caused vampire traits, and even *The Vampire Lestat,* who traced his vampire genetics back to a chance combination of different species, have at least some scientific credibility.

The notion of vampires biting humans to transform them into more vampires, or a separate species coexisting alongside us, at least presents some interesting ideas as to how it might happen at a genetic level. The sad truth is that it would take an extraordinary number of coincidences for any of this to occur in reality. The chances of being able to turn a human into a vampire by tinkering with their genetics is a long way off, if it is even possible at all. For the time being, at least those who hanker after the vampire lifestyle will have to make do with cosmetic changes, and that is the subject of the next chapter.

CHAPTER 10

Vampiroids

> 'M. X affirms ... She's nothing more than a decomposing body, and has no need to eat for living, she cannot die a natural death, she exists eternally if she's not burned,'
>
> Jules Cotard, *Du délire hypocondriaque dans une forme grave de la mélancholie anxieuse*

Poor Renfield, the unloved lunatic in Dr Seward's sanitorium. All he wanted was for his master to love him. He tried everything, from praise to carrying out favours and mimicking his master's way of life, but all to no avail. When Oscar Wilde said that 'imitation is the sincerest form of flattery that mediocrity can pay to greatness,' he probably wasn't thinking about Renfield locked in a madhouse eating flies and spiders.

Renfield was very much overlooked in *Dracula*. The Count had no real interest in him other than using him to gain access to more victims. The other characters in the book scarcely noticed him; even his doctor, Jonathan Seward, is only interested in his patient's strange behaviour and quickly forgets about him when something more interesting happens. Even when Renfield attacks Seward with a knife, cutting his arm and licking up the blood that drips onto the floor, the doctor sees it as a strange but isolated incident. It is only after Renfield has been killed by his idol, the Count, that Seward even realises the connection between the two.

Vampirology: The Science of Horror's Most Famous Fiend
By Kathryn Harkup

Published by the Royal Society of Chemistry, www.rsc.org

Though Renfield is a relatively minor character in *Dracula,* he has had a huge influence on the field of psychology, even having a syndrome named after him. He was the original 'vampiroid', a person who aspires to be a vampire. Such aspirations can take many forms; by adopting a vampire aesthetic, or by behaving in a way that is often characterised as vampiric. Renfield was clearly at one of the more extreme ends of the vampiroid spectrum. Though his behaviour mimicked that of his hero, he made no attempt to look like Dracula. Renfield was interested in blood, not aesthetics. And, judging by Stoker's descriptions of Dracula, the Count didn't cut such a handsome figure that others would want to imitate him. The female vampires in the novel, however, were repeatedly described in terms of their beauty and attractiveness.

Getting the vampire look and adopting vampiric behaviour are related, but different things. Since the times of Lamia in ancient Greece, blood-sucking beings have used their appearance to attract victims. Admittedly, vampires let themselves go a little during the seventeenth and eighteenth centuries, when they were characterised by their dishevelled and swollen appearance. But, in the nineteenth century, fictional vampires realised the benefits of looking good and made the most of it. Since then, vampires have rarely looked back. Modern vampires are generally accepted to be good looking, at least long enough to get close to their victims (faces can sometimes change when a human is in a vampire's clutches; see Chapter 7). They also often have a sharp eye for fashion, usually in darker-coloured fabrics.

Most people, even those who haven't read vampire novels or watched vampire films, would be able to list a few distinctive features that make vampires easily identifiable. This visual shorthand for the blood-sucking undead is a relatively recent phenomena and is by no means uniform. In the past, male and female vampires were often strikingly different in appearance, particularly in the nineteenth century. Both served as warnings, but in a very different way.

10.1 THE FEMALE OF THE SPECIES IS MORE DEADLY THAN THE MALE

Victorian men generally preferred their women to be passive. There was a popular artistic theme at that time of women asleep, ill or dead. An early example, with prominent Gothic or vampire imagery, can be seen in Henry Fuseli's 1781 painting, 'The Nightmare', in which a woman is laying unconscious, or possibly dead, with a demon-like creature sitting on top of her. Strong, assertive women were also popular themes, but they were usually portrayed very differently – in predatory roles, such as Medusa or vampires. Edvard Munch painted six different versions of 'Love and Pain', also known as 'The Vampire' between 1893 and 1895. Another painting, titled 'The Vampire' was produced by Philip Burne-Jones in 1897, the year *Dracula* was published. Burne-Jones' version depicts a woman leaning over an unconscious man. The model is believed to have been the actress Mrs Patrick Campbell, and there were rumours that she and Burne-Jones had an affair. The painting perhaps gives an insight into how he felt about the relationship.

Burne-Jones' painting was inspired, in part by Fuseli's 'The Nightmare', but perhaps also by another image, used to illustrate Sheridan Le Fanu's vampire novel, *Carmilla* (see Figure 10.1). Le Fanu's story can be read as a warning against the dangerous seductive qualities of beautiful women with strong independent personalities, something that Victorian men seem to have been particularly fearful of. The only Victorian women who were expected to show any interest in sex were prostitutes, who were associated not only with sin but also with the risk of venereal disease. Attractive female vampires were therefore yet another veiled reference to syphilis (see Chapter 8).

Bram Stoker continued the theme of the fatal female in *Dracula,* and cast his female vampires from the same mould as Le Fanu's Carmilla. Stoker showed Lucy Westenra's transformation into a vampire made her more physically attractive. She became more dangerous as she became more beautiful, and is even nicknamed the 'bloofer (meaning beautiful) lady' by the children she preys on in her vampiric state. Dracula's three vampire brides are so captivatingly attractive, and therefore dangerous, that even the Count himself seems to fear them, and protects Jonathan Harker from their bloody attentions.

Published the same year as *Dracula*, Florence Marryat's novel *The Blood of the Vampire*, had a female vampire lead, Harriet

(a) (b) (c)

Figure 10.1 (a) Thomas Burke's 1783 engraving of 'The Nightmare' by Henry Fuseli; (b) David Henry Friston's illustration for *Carmilla* by Sheridan Le Fanu; (c) 'The Vampire' by Philip Burne-Jones.

Brandt. Harriet had 'a far more dangerous quality than that of mere regular feature. She attracts without knowing it. She is a mass of magnetism.' Anyone drawn into her sphere of influence was drained of vitality to the point of death. Harriet herself was innocent of any malicious feeling towards these people, but Marryat managed to include almost every social prejudice of the age to show that her vivacious and charming vampire was inherently dangerous. Harriet's parents were murderers who delighted in torture and, to make matters even worse, they had never gone through the formality of getting married. In Marryat's world, the combination of an illegitimate birth, a mother of mixed-race heritage, daring to be a woman, and attractive to both men and women, made Harriet irredeemably evil.

The warnings were stark, but women could sometimes use this imagery to their advantage. Sarah Bernhardt, the famous actress of the late nineteenth and early twentieth century was famed for eccentricity, as well as her abilities on stage and screen. She tapped into the imagery of dead or unconscious women for her benefit. One of the most famous photographs of her, from 1873, shows her sleeping in a coffin. This was something she apparently did regularly and also relaxed in it when studying for a role. Thousands of copies of the snap were sold as a postcard to her adoring fans.

Other actresses used vampire roles to achieve fame and fortune. Theda Bara colluded with her film studio to create a mysterious and exotic personal back-story to promote her career in the early days of the silver screen. She adopted a distinctive makeup style to make sure her features showed prominently under the harsh lighting needed to record images on film at the time. Film cameras back then did not respond to every colour in the same way. Reds showed up as dark black, but blues appeared white. Bara feared her blue eyes wouldn't be visible on screen, so did everything she could to make them stand out. In real life, her distinctive eye makeup would have been produced using red and coral shades, but when translated to black and white film it appeared as though she had been very liberal with black khol eyeliner.

Bara rose to fame in the 1915 film, *A Fool There Was,* in which she played 'The Vampire' – a woman who was very much alive and who seduced, drained and ruined the men she encountered, several of whom die over the course of the film. There was no blood-sucking involved, but it was very much in keeping with nineteenth-century images of female vampires: they were beautiful but lethal.

Cinema, even from the very beginning, has had a huge influence on style and fashion. The big screen has the ability to make almost anything look glamorous, even vampires. Bara was one of the first film stars, and was often typecast in 'vamp' (short for vampire) roles. Despite playing very unpleasant characters, women wanted to look like her and she was a major influence, along with fellow actors Louise Brooks and Pola Negri, on the 'vamp' look that was popular in the 1910s and 1920s.

By the early twentieth century, female vampires had become fashion icons, but the men were lagging behind on the desirability stakes. Victorian authors still gave warnings about the dangers of male vampires *via* their appearance, but in a very different way.

10.2 JUDGING BY APPEARANCES

The modern male vampire aesthetic has its origins in John Polidori's short story, 'The Vampyre'. The pale, suave, well-dressed Lord Ruthven of the tale has inspired many imitators. At a time when a person's physical appearance was thought to tell you a lot about their character, the physical description of characters in novels and short stories told you a lot about what to expect of their behaviour. Lord Ruthven was a thinly veiled Lord Byron and shared many physical similarities with the notorious aristocrat, known for his immoral behaviour. Therefore, Ruthven's appearance signalled that Polidori's character might also be 'mad, bad and dangerous to know'. Later nineteenth-century fictional vampires are all direct descendants of Lord Ruthven, and they inherited his paleness, glinting eyes and aristocratic position in society. But beauty and danger were not generally equated in the same way for men as they were for women. Polidori excepted, Bram Stoker, and other vampire writers of the time, portrayed their male vampires very differently, as typified by James Malcolm Rymer.

In *Varney the Vampire*, Rymer's epic vampire tale from the middle of the nineteenth century, Varney was tall, gaunt and with eyes of a 'savage and remarkable lustre'. There was an explanation for his remarkable features – they had apparently been distorted as a result of spasmodic contractions of the face muscles as he was being hanged. Galvanism had managed to restore his life, but not his looks. Clearly, such a remarkable-looking individual was not to be trusted. With such an unappealing cadaverous appearance, he had to rely on his wealth and stealth to attract and gain access to his victims.

The appearance of male vampires during the nineteenth century, particularly in *Dracula*, played up to psuedo-scientific theories of the time that suggested physical appearance gave clues to personality. There is a long history of people associating physical characteristics with behavioural traits, but these ideas started to be collected together and investigated as a unified theory in the late seventeenth century. The Enlightenment had opened up all aspects of the universe and its contents for investigation by natural philosophers, and human behaviour was not to be left out. During the nineteenth century, phrenology was established as the study of the skull for clues to personality. The theory ran that the brain was the organ of thought and was clearly not uniform. Different areas of the brain were responsible for different mental activities, and some might be enlarged as they became more developed. This underlying brain shape was thought to be transferred to the skull as it solidified round the brain of a growing infant. Reading the bumps on the head was believed to give indications as to whether your child was to grow up to be a musician or a mass murderer.[†]

The theory of phrenology developed and incorporated ideas from other branches of science and pseudo-science including mesmerism, evolution and anthropology. Phrenology, in turn, influenced ideas about education and psychiatry, and was employed as justification for racism. What may have started off as an interesting idea has since been firmly established as pseudo-scientific nonsense, but sadly these ideas persisted and grew throughout the nineteenth century.

Criminology also came under the influence of phrenology. Cesare Lambroso, an Italian anthropologist, was convinced that some people were born criminals and could be identified as such by their bulging brows, bushy eyebrows, ape-like noses, disproportionally long arms and large jaws. These traits he saw as throwbacks to a more primitive species. His theories were unfortunately very influential. First published in Italian in 1876 as *L'Uomo Deliquente*, and with an English translation appearing in 1895 under the title *Degeneration*, these ideas became popular throughout Europe. Traces of the theory can be found in many

[†]Though these characteristics may not be mutually exclusive.

literary works of the time, but it is more prominent in some than others.

In Oscar Wilde's 1890 novel, *The Picture of Dorian Gray*, no one could believe that Dorian had led the debauched and evil life that was constantly rumoured about him. But, rather than Dorian's own face showing the clear signs of his sinful lifestyle, it was a portrait in his attic that took on these characteristics. Dorian maintained his youthful beauty for years after they should have faded, an attribute that links the novel to the vampire genre.

In *Dracula*, Bram Stoker's description of the Count's physical appearance would clearly indicate him as a criminal for anyone familiar with Lambroso's theories. The many descriptions of Dracula's appearance in the novel – the long arms, bulging forehead, ears pointed at the top, bushy eyebrows *etc.* – were examples of features that were believed at the time to indicate criminal behaviour. He was never described as handsome or attractive and had to use his mesmeric abilities to attract and overpower his victims.

It would be a struggle to describe male, nineteenth-century vampires as handsome.[‡] Distinctive might be a better word because, regardless of how appealing you may find them, they were certainly easy to identify. Ask anyone today to describe a vampire and they are more than likely to pick out the key features that were established two hundred years ago – pale and cadaverous. And, as with their female counterparts, it was popular culture that helped establish those features and eventually transform them into something appealing and desirable that people would want to emulate.

10.3 UNUSUAL MATINEE IDOLS

The appearance of nineteenth-century male vampires clearly identified them as criminals, but they did a considerably better job at blending into society than the red, bloated, flaking vampires of folklore. Vampires started to tidy up their appearance for a night out at the theatre back in the 1820s.

[‡]Each to their own, of course. Who am I to judge?

It was less than a year after Polidori's 'The Vampyre' was published that it was adapted for the stage. The renowned actor Thomas Cooke[§] appeared as Lord Ruthven at London's Lyceum Theatre, the future workplace of Bram Stoker, in a play titled *The Vampire, or The Bride of the Isles*. Polidori's tale of the aristocratic vampire went on to inspire a string of theatrical productions. London theatres were soon overrun with vampires, but they were not quite the sophisticated, elegant creatures we know today. Many of these early theatre productions were closer to pantomime than serious plays. There were comic moments, topical gags and knockabout farce, but they were extremely popular and were still being staged over fifty years later. The craze even spread to France, where one theatre critic lamented, 'There is not a theatre in Paris without its vampire!'

Although vampires enjoyed a prolonged popularity on the Victorian stage, they were not yet characterised by a particular style. However, the first vampire theatrical production did introduce one important feature – the vampire trap. This was a spring-controlled trapdoor that allowed the actor playing the undead role to vanish from the stage in the blink of eye. These sudden disappearances frightened and delighted audiences, and the vampire trap became a mainstay of stagecraft design.

Fashions fade and vampires eventually lost their appeal for Victorian theatregoers. When Bram Stoker published *Dracula*, though he was extremely influential in the theatre world, he only ever staged one theatrical reading of his work. The audience was mostly made up of cleaners and back-stage staff who had nothing better to do at the time. It was not even close to a full-scale production and was simply carried out to protect the author's copyright. The work was granted the Theatrical Licence Number 162 and promptly shelved. Though Stoker would have loved for his boss, the renowned actor Sir Henry Irving, to take on the title role, he refused to even consider it. Irving dismissed the work, perhaps noting some uncomfortable similarities between himself and the malevolent vampire.

[§]Cooke would later make a name for himself playing the nameless creature in theatrical adaptations of *Frankenstein*, the other Gothic horror classic produced by Byron's storytelling challenge at Lake Geneva back in 1816 (see Chapter 1).

Vampires made a triumphant return to the stage in 1927. The actor Hamilton Deane approached Bram Stoker's widow for the rights to stage her late husband's novel and gained the official stamp of approval. Deane wrote the script himself and condensing the novel into three acts performed on a single set. Though scorned by the critics the play was a huge success. It brought Florence Stoker some much-needed cash and it was this production that established the vampire look for the rest of the century.

Deane set the entire play in the drawing room of a posh London house, which meant the Count would only ever be seen in the evening and therefore dressed in fashionable evening attire – an evening suit with white tie. Vampires had shrugged off their folklore past, survived ridicule in London's vaudeville, and finally been fully accepted among the fashionable upper classes. Blending into polite society was one thing, but it could be taken too far.

Audiences had to be able to clearly distinguish the character of the vampire from all the other men on stage. When all the men would have been wearing near identical dinner jackets and ties, this was not so easy to achieve. So, the Count was given a long opera cape with a high collar. It's distinctive outline made the character easy to identify on stage. The cape gave the character a striking bat-like silhouette and the actor wearing it could peer over the top of the collar in a sinister manner whenever the script demanded. It was also a useful way to make a dramatic exit. The cape, held by fellow actors, masked the vampire from the audience as he dropped through the stage trapdoor. Subsequently, opera capes with a high collar would forever be associated with vampires.

Deane had taken a big risk with *Dracula,* but it paid off. Florence Stoker had needed some considerable persuading to give her permission for the production. Her caution was understandable. She had learnt how tenaciously she had to protect her means of support, and not just from theatres and playwrights wanting to stage adaptations of *Dracula*. By the time Deane was planning to stage the play, theatres were competing for audiences with cinemas. The silver screen, with its sharp black-and-white contrast, was the perfect medium for vampires, and they flourished there.

10.4 FILMS, FANGS, FAME AND FORTUNE

Vampires lent themselves extremely well to early cinema. Even the best cameras at the time were poor quality by modern standards and struggled to record images with much clarity. Strong lights were required to illuminate everything, so the reflections were bright enough to make an impression on the film, and this tended to bleach out the actor's face. The result of strong lighting and low definition was a dramatic contrast between bright white faces and deep, dark shadows – perfect for creepy vampiric effects and moods.

The first vampire film was of course F. W. Murnau's 1922 classic *Nosferatu: A Symphony of Horror*. It was adapted from *Dracula,* but with only cursory changes to avoid copyright infringements. These small changes had not been enough for Florence Stoker, who demanded £5000 in royalties and the destruction of all copies of the unauthorised film. Her lengthy and bitter battle was largely successful in that *Nosferatu* rarely appeared on cinema screens for much of the twentieth century. It is perhaps for this reason that *Nosferatu*'s vampire, Count Orlok, is something of an outlier when it comes to the vampire look.

Murnau's Orlok was created before Deane's theatrical interpretation that had done so much to cement what a vampire should look like in the public mind. This first film vampire was much more closely related to the Dracula of the book than the vampires of the stage. He was tall and thin but hunched over, he had a bald head, pointy hairy ears and rat-like teeth, and unfeasibly long fingers that ended in vicious-looking nails. He dressed shabbily in a long tatty coat, with no sign of a cape or even a dinner jacket.

Count Orlok and *Nosferatu* were completely overshadowed for much of the twentieth century by the 1931 film *Dracula*, produced by Universal Pictures. The film company paid Florence Stoker £40 000 for the rights to the novel. They also bought the print of *Nosferatu* from the original German production company for £400 and promised Florence they would destroy it. Fortunately, they neglected to incinerate what is now recognised as a landmark in film-making.

The 1931 film, with Bela Lugosi as Dracula picked up from where the Hamilton Deane theatrical production left off.

The screenplay was largely based on Deane's authorised theatrical adaptation. Cinema had the ability to reach much further across the globe and play to bigger audiences than could ever be achieved by a theatrical production, no matter how popular it was and for however many years it toured. The long opera cape with its high collar became the standard attire for any vampire. From that point on a dinner suit and an opera cape was enough to identify any character as a vampire, from a cartoon duck to a purple Muppet singing about numbers. The association for Bela Lugosi was so strong that he is barely known for any of his acting work outside of vampire roles. He fully embraced his iconic status and was buried in his Dracula outfit after his death in 1956.

The only thing that Lugosi's look didn't have was fangs – these were introduced as a staple in the 1950s Hammer Horror films. Though *Dracula*, and other vampire novels before it, made much of the sharp, pointed teeth of vampires, they were not always a feature of their folklore predecessors. Different vampire-like creatures across Europe and the rest of the world have found different means to suck the blood of their victims. There are the Filipino *aswangs,* which have a sharp hollow tongue that they use to pierce the skin and suck up the blood, similar to Bulgarian vampires who have a sting or barb at the end of their tongues. Other vampiric creatures also bit arms, or the chest around the nipple or heart, but rarely the neck. The combination of sharp teeth and neck-biting was made by filmmakers because of its sex appeal rather than any benefits to blood-drinking.

The Bela Lugosi vampire look became almost a uniform for any vampire character in a film and it scarcely changed through the many reincarnations of Dracula and other vampires by fantastic actors such as Peter Cushing, Vincent Price and Christopher Lee. In fact, like Lugosi before them, it is through these roles that many of these actors achieved their fame. Not only did the Hammer Horror films adopt Dracula's uniform from the 1931 film, Christopher Lee wore the same ring as Bela Lugosi in all his vampire roles. Only in more recent years have film vampires managed to shed their antiquated dress sense and adopt modern garb. Even in modern clothes, vampires usually retain some of the characteristics from their origins back in the early nineteenth century – dark, well-cut clothes, pale faces and blood-red mouths.

The cinema had quickly discarded Stoker's original descriptions of the Count in favour of a more glamorous look. Bela Lugosi's iconic portrayal was filmed when Lambroso's theories of biological criminality were still worryingly current. His arguments had no basis in fact, but they did have enormous influence in the early twentieth century; from penal systems to Mussolini's fascist policies. Even with the benefit of close-ups, the cinema dispensed with the specific facial characteristics that Lambroso would have identified as criminal. Instead, films used dramatic lighting and music as cues for the audience to understand that vampires were the bad guys.

As time went on, vampires became more and more handsome, reaching absurd levels in the Twilight series, where almost every page, and sometimes it feels every paragraph, asserts just how extraordinarily beautiful all the vampires were. Living in parallel with the most beautiful specimens of vampire-hood were their less attractive cousins. The rediscovery of Murnau's *Nosferatu* brought an alternative view of vampires to the fore. Later twentieth-century vampire incarnations, for example, in *Buffy the Vampire Slayer*, show Count Orlok's influence. Many vampire films, such as *What We Do in the Shadows* and *Shadow of the Vampire*, now directly reference the 1922 film.

Lambroso's outdated theories have been long disproved and largely forgotten, but there is still an unpleasant lingering aftertaste of these ideas even today. Though the specifics of the theory have been forgotten, the general idea is depressingly persistent. Depictions of vampires in the twentieth century fall into two broad categories: very human and very attractive, or those with hideous demonic or distorted features. Those that are ugly are quickly dispatched without a second thought, as they are clearly evil and deserve to be terminated. Attractive vampires quite literally get away with murder. Even when they are predatory, remorseless killers, they seem to be forgiven and even idolised.

The idea of an attractive vampire may have started in the early nineteenth century, but the glamour of the stage and silver screen has only added to the vampire's appeal over the decades. With its glamorous associations, many people have imitated the vampire look, a fashion trend that has continued long after the 1920s vamps. Many people outside of film and theatres feel happiest dressed in the clothes and general look associated with vampires. Much of this is achieved with make-up and shopping in stores that cater for these fashions, or individuals may tailor or adapt their own clothes. Fangs, a vampire staple of the nineteenth century literature and most films, are an important feature needed to look the part of the vampire. Some go beyond plastic joke shop versions that slip over their own teeth and find obliging dentists who will remodel their incisors into a more pointed shape.

There are some who have adapted their living arrangements, not just their appearance, to become more vampire-like. The nineteenth century actress Sarah Bernhardt was not the only one with unusual sleeping arrangements. Around the same time there was Eric Stanislaus Stenbock, Count of Bogesund in Sweden and heir to the estate of Kolga in Estonia,[¶] who lived most of his life in England. Stenbock was the author of *True Story of a Vampire*, and was also known for sleeping in a coffin. He cultivated a range of eccentricities, including keeping an extensive menagerie of exotic and unusual pets; one of which, a toad, always sat on his shoulder at meals. Stenbock could pick and choose out of the many vampire characteristics those that appealed to him, and he had the money and status that allowed him to indulge himself.

There are certainly vampire characteristics with potentially broad appeal. Through most of the nineteenth and twentieth centuries they have been portrayed as beautiful and able to maintain their youthful good looks for eternity. Who wouldn't want that? Some people have set up organisations to exploit exactly those desirable qualities of vampirism. The Temple of the Vampire sees vampires as the next step in human evolution (see Chapter 9) and offers its followers a guide to achieving a better life through worldly success, physical immortality and personal happiness. There are meetups, conclaves and guides to achieving these goals. Fewer people seem to want the other characteristics

[¶]Stenbock inhereted the Kolga estate in 1885 from his uncle, Count Magnus, coincidently (or not) the name of the vampiric character in M. R. James's short story published in 1904.

that go along with the more appealing vampire traits – like being undead, or the blood-drinking – but it takes all sorts.

10.5 THE WALKING DEAD

Believing you are dead is irrational. Simply being able to vocalise such a belief would be taken by most people as clear evidence that a person was alive, but it does happen. There is a very rare psychiatric disorder where patients develop the belief that they are dead or missing body parts. The condition is known as Cotard's syndrome, after the neurologist Jules Cotard who first described it.

Symptoms of Cotard's syndrome can vary. Some claim that all their blood has been taken, while others fear they are missing their heart or brain. Around half of the patients in a 100-patient survey denied they existed, while the other half claimed immortality. An important step in developing the condition appears to be depersonalization, or loss of a sense of self. If a patient's own body feels unrelated to them or alien in some way, being dead is a way of explaining their body's unfamiliarity. Understandably, patients are usually very anxious and depressed, sometimes to the point of suicide. The syndrome is most often associated with underlying medical conditions, but there is a long list of them: from depression to typhoid fever, and from stroke to taking certain antiviral drugs. Depending on the underlying disorder, this altered state can persist for weeks or even years.

Though the brains of many sufferers of Cotard's syndrome have been studied using a number of techniques, there is no clear single difference between their brains and anyone else's. The condition doesn't even have its own official classification and is often treated as a symptom of an underlying condition. Treatment, therefore, is usually based on the believed cause of Cotard's syndrome and can vary from simply removing the cause of the problem, such as stopping the anti-viral drugs, or using drugs and therapies commonly applied to psychological disorders, such as anti-psychotic drugs or electroconvulsion therapy. Sometimes recovery is spontaneous.

Though the links between being dead and immortality also have strong associations with vampires, sufferers of Cotard's syndrome do not believe they are vampires. They display no other vampire characteristics such as aversion to sunlight, or a desire to drink blood, even if they believe they have lost their own. There has only been one recorded case of Cotard's syndrome being linked to anything even close to vampirism, and that was a man who simultaneously believed he was dead and had turned into a dog. He was diagnosed as suffering from both Cotard's syndrome and lycanthropy – a word linked to werewolves, but also a talent for transformation displayed by Dracula (see Chapter 7).

People suffering from Cotard's syndrome do not seek it out as a lifestyle choice. They do not believe they are dead because it is something they consider desirable. It is a very distressing condition, which sufferers seek to remedy. This is not necessarily true of those who display another vampire characteristic – the desire to drink blood.

10.6 SANGUINARIANS

There are people who enjoy drinking human blood or having their blood drunk by someone else. In the past there have been stories of venues in New York where patrons would drape themselves over the bar with a cannula fitted ready to be fed on by other customers. The arrival of HIV and AIDS in the late 1970s and early 1980s put a stop to these kinds of places. Instead of feeding directly from the source, there are now bars and restaurants with blood-based menus. Sanguinarians also generally pair up in private, one donor and one drinker. There are networks to support wider communities of blood drinkers. They share information about safe practice and search for scientific research to explain their thirst. What they don't all do is dress up as vampires. Though vampire look-a-likes are not necessarily excluded from their groups, it is not a requirement.

The amount of blood that is drunk varies from 20 ml every couple of weeks to 300 ml or more. They say the blood makes them feel better and they feel unwell if they abstain for lengthy periods. Some complain of headaches, tiredness and sensitivity to sunlight. This sounds as though there is something in the blood that is correcting a deficiency, or activating feelings of well-being. There is nothing specific to human blood, that I am aware of, that could cause any of this.

Protein, vitamin or micronutrient deficiencies can be treated without drinking blood and there is no compound in blood that would give the drinker a sense of euphoria. It is true there are some chemicals that when introduced into the body can find their way into the brain. There they can interact with receptors in the ventral tegmental area (VTA) and give rise to feelings of pleasure. The brain naturally produces dopamine that stimulates the VGA as part of the body's reward pathway that encourages us to do things that will help us survive and reproduce. For example, this is why seasoned food tastes better; our body is rewarding us for eating salt that it needs to function properly. Other things, like drinking water when thirsty, or sexual contact, likewise trigger the release of dopamine and good feelings. Some compounds can increase the release of dopamine or keep it circulating for longer. For example, cocaine stops cells from reabsorbing dopamine and nicotine makes cells release their dopamine more easily. However, there is no compound in blood that could cause a chemical addiction in this way, so far as I know.

However, addiction is a complicated thing, and not just a straightforward biological response in our body. There are cultural, social and habitual aspects to addiction that are difficult to separate out. And there is a spectrum of responses to addiction, from those who struggle to kick a forty-a-day smoking habit, to those who only light up on a night out. Persuasion can be very powerful too. People who believe a particular pill will make them well can report feeling better even when they have swallowed a placebo. Just as there is a spectrum of responses to substances, there is also a wide spectrum of blood drinkers and many names for the different points along that spectrum.

Clinical vampirism is a term sometimes used for those who have a compulsion to drink blood. It is more commonly called Renfield syndrome, after the character in *Dracula* who fed his blood needs by eating flies, spiders and other small animals. He was compelled to act in this way because he believed 'the blood is the life' and aspired to be more like his hero Count Dracula.[||]

[||]As well as books on psychiatric disorders, Bram Stoker also consulted his brother Thornley Stoker, who was a highly respected doctor and president of the Royal College of Surgeons in Ireland.

Whatever it is called, very few clinical cases have been described. For some, drinking blood, their own or someone else's, as well as bleeding themselves, has a sexual element, known as haematomania, or a blood fetish. At one end of the spectrum are sanguinarians who pair up for a mutually beneficial blood drinking** and at the other extreme are the few who have committed murder to obtain blood.

There are very few examples of serial killers who have been motivated by blood, but the few that there have been received a disproportionate amount of press attention. The press was also keen to apply phrases like 'real-life vampire' in its reporting to sensationalise these crimes. Given how most vampires throughout history have been seen to cause the death of humans, and the notable loss of blood in their victims, it's not exactly a difficult link to make. But those connections go both ways.

10.7 CRIMES AND MISDEMEANORS

That a compulsion to drink human blood could lead someone to murder in order to satisfy their need is extreme and, consequently, very rare. However, several real-life serial killers, apparently motivated by blood, have been very closely associated with vampires. And, real-life serial killers with a lust for blood may have inspired some aspects of vampire literature in the nineteenth century. Sabine Baring-Gould collected together an astonishing list of killers who had a taste for human flesh and blood in *The Book of Were-wolves*. Published in 1865, the book included lurid details of the Blood Countess who we met in Chapter 4. The details of Erzsébet Báthory's alleged crimes included in this book may have been Stoker's inspiration for the vampire brides in *Dracula*, at least in part.

Stoker may also have been inspired by perhaps the most famous serial killer of all time. A couple of years before he started the research on his vampire novel, there was a vicious murderer prowling the streets of the Whitechapel district of London. The identity of the person who carried out the brutal murders of five women, who were killed and then horribly

---

**Some sanguinarians, apart from human blood, are otherwise vegetarian in their diet.

mutilated, is still unknown. This unidentified serial killer became known as Jack the Ripper. Though *Dracula* makes no direct allusion to the Whitechapel murders, and the press at the time made no connections between the killer and vampires, it may have played on his readers' minds when reading about the Count prowling around London in search of women victims. The publication of *Dracula*, and its huge popularity, also made it easier for journalists to describe brutal murderers as vampires. Two of the most famous 'vampire' serial killers were Fritz Haarman and Peter Kurten.

In the mid-1920s in Hanover, Germany, police discovered human bones dredged from the river Leine. Whoever the bones had belonged to had died recently, and the hunt was on for what the press called the 'Vampire of Hanover' or the 'Werewolf of Hanover'. The police eventually arrested Fritz Haarmann and his accomplice, Hans Grans, but not before they had lured dozens of men to their death. Grans had picked up these young men and invited them back to his squalid flat. There, they would be fed until they were sleepy and then Haarmann would force himself on them, gnawing through their neck during sex. The remains of his victims were butchered and some parts were eaten. Meat and clothes were sold at the city's markets. Offal and bones were discarded in the river. Haarmann had killed at least 27 young men, and possibly up to 50, before he was caught. In April 1925, at the culmination of a sensational two-week trial, Haarmann was found guilty and sentenced to death. He went to the scaffold joyfully.

Five years later, in Dusseldorf, numerous bodies of women and girls were found – slashed and sometimes decapitated by knives and scissors or bludgeoned with hammers. This time, the hunt was on for the 'Dusseldorf Vampire'. Peter Kürten confessed to his wife that he was the 'Vampire of Dusseldorf' and urged her to turn him in so she could collect the reward money that was offered. Upon his arrest, he immediately confessed to everything the police accused him of, plus two other unsolved murders. He admitted he had drunk the blood of his victims because he found it sexually exciting.

Though he was believed to have been responsible for ten murders and 31 attempted murders, he was tried for nine murders and seven attempted murders. He was found guilty on all counts and sentenced to death by guillotine. Before he placed his head on the block, he apparently asked if he would remain conscious long enough after his beheading to hear the blood gushing from his neck – the idea excited him. Rather than a punishment, his execution was something Kürten relished.

Such unusually brutal crimes came under considerable scrutiny from many quarters. Apart from the sensational accounts and continued references to vampires in the press, there were also serious studies into what could have motivated these men. Both men were subjected to psychological examinations. Haarmann declared he did not know why he killed, but denied that he was mad. Doctors confirmed he was competent to stand trial. Kürten was also examined and several psychologists pronounced him completely sane and fully aware of the seriousness of his crimes.

Haarmann and Kürten never seem to have associated themselves with vampirism, that was a label the press gave them. However, some serial killers, have tried to use the vampire link to their advantage. John Haig, the acid bath murderer, so-called because he dissolved the bodies of his victims in drums of acid, claimed to be a vampire when he was questioned by police. Many believe this was a deliberate ploy on his part to be declared insane so he would not face the death penalty. However, he had shown no signs of killing his victims to obtain their blood, it seems their money was what he was after. His claims of blood lust were not believed, and he was convicted of killing six people (though he claimed there were nine) and sentenced to hang on 24 July 1949.

In 1959, there was another case of a criminal deliberately associating himself with vampires. The then-sixteen-year-old Salvador Agron killed two people while dressed in a long black cape with a red lining, clearly in imitation of Bela Lugosi's Dracula character. His distinctive dress led him to being called 'The Capeman'. His choice of weapon, the sharpened tip of an umbrella, meant he was also known as 'The Umbrella Man'. Agron was head of a New York gang that called themselves 'The Vampires'. He and his fellow gang members were on their way to a 'rumble' or street fight with rival gang, the Norsemen, when they came across two teenagers. They mistook the teenagers for members of the Norsemen and Agron stabbed them to death before fleeing. He was eventually caught and convicted.

Murders and murderers have always held a macabre fascination for the general public. Newspapers and publishers of true crime books thrive off it, and the more unusual or brutal the killings, the greater attraction they seem to have. Anything with a vampire theme is bound to sell extra copies and this obsession with the occult and death reached fever pitch in the 1970s with stories of the Highgate Vampire.

In 1969, Bishop Sean Manchester claimed to have found evidence of vampire attacks in and around London's famous Highgate Cemetery. David Farrant claimed to have seen a supernatural figure when he passed by the cemetery one evening in 1970. Soon, more reports emerged of ghostly figures in and around the cemetery, including a tall man in a hat, a spectral cyclist and a woman in white. The media lapped it up, especially as Farrant and Manchester developed a bitter rivalry.

Over the following years, graves were disturbed and bodies desecrated in search of the alleged vampires. Manchester claimed he had brought the vampire infestation to an end in 1974, when he discovered a coffin containing the head vampire in an abandoned Gothic house in the Highgate area. He said he had cremated the remains in the garden of the house. Eventually, the media lost interest and things returned to normal in the area. However, myths and rumours surrounding the Highgate vampire persisted, no doubt fuelled by the ongoing acrimonious battle between Farrant and Manchester that only ended with Manchester's death in 2019.

A few acts of vandalism and some good storytelling developed into a huge story for the press. No one had died and no firm conclusions about what really went on were ever reached. But, the Highgate vampire has captured the public imagination for decades, showing just how influential vampires can be, and how little has changed since the tales of vampire epidemics in the eighteenth century.

10.8 I WANNA BE LIKE YOU

Two hundred years ago, no one would have wanted to look like a blood-sucking murderous vampire. Today, it is fun for all the family. No parent would think twice about dressing their child in a cape, applying a little make-up to make them look pale and handing them some plastic fangs. But, when you think about it, dressing children as predatory blood-suckers is a bit weird.

It is a considerable personal disappointment that more people do not seek to imitate the Albanian vampire-like creature, the *sampiro*. If anyone is looking for a Halloween costume for next year, these versions of the undead are great inspiration. These revenants are swathed head-to-foot in floaty, diaphanous fabric, through which their huge eyes glow like car headlights in a fog. They clatter down streets after their prey, swaying unsteadily in high heels and making kissing noises as the smack their lips in anticipation of their blood meal.

The blood-sucking undead come in many forms across the world, and they have been through many transformations over the centuries, but there is one that has come to dominate. Of all the variations, it is the pale aristocratic vampire of the Romantic era that has prevailed in terms of a desirable vampire look. The pale, gaunt vampire of the early nineteenth century has become the standard. The influence that films, and other forms of mass media, have on our lives is incredible and pervasive.

Most people have managed to mentally disengage the vampire aesthetic from the undead's murderous behaviour. Those who wish to look like the pop-culture version of vampires are imitating a fashion style, rather than vampire behaviour, though some may adopt vampire traits they find appealing. Sleeping in coffins may not be standard practice, but it might not be as uncomfortable as it looks. They are made to measure after all. Even those who drink human blood find consenting partners.[††] Their way of life may not be typical but it does no harm to anyone else. Live and let live. Or, live and leave the wannabe undead to their own devices. It is only in exceptionally rare cases that things have spilled over into criminal activity, but these few individuals have been responsible for a lot of bad press.

Hopefully nothing in this book, so far, has left you worried or concerned that the undead may be out there and planning on drinking your blood. But, just in case any doubts have started to creep into you mind, fear not. The next chapter is all about how to protect yourself from blood-sucking revenants.

[††]The important word here is 'consenting'. I hope this only needs emphasising and not explaining.

CHAPTER 11

Prevention

> '"Will your ladyships be pleased to buy an amulet against the oupire, going like the wolf, I hear, through these woods," he said dropping his hat on the pavement. "They are dying of it right and left and here is a charm that never fails; only pinned to the pillow, and you may laugh in his face."'
>
> Sheridan Le Fanu, *Carmilla*

On 1 January 1701, the islanders of Mykonos celebrated finally ridding themselves of a very troublesome vampire. Their failure to spot and properly deal with this vexatious revenant had led to weeks of misery and upset. They should have known that this quarrelsome individual would continue to harass the locals even after his death, but they had evidently failed to take the proper precautions.

According to the local traditions, bodies should have been buried swiftly after death. One commentator wrote that they usually scarcely left enough time for the body to cool before it was interred. Unfortunately, in this case, the undead peasant who had become so bothersome had been found murdered in a field; exactly when and by whom was unknown. Though he was swiftly buried, it was perhaps not quick enough. His unpopularity in life may have led to some of the elaborate details expected around burials and mourning for the dead being skipped over. The result was that, two days after his funeral, he

Vampirology: The Science of Horror's Most Famous Fiend
By Kathryn Harkup

Published by the Royal Society of Chemistry, www.rsc.org

had become a *vrykolakas*, one who returned from the dead to inflict misery on the living.

Fortunately, the newly formed *vrykolakas* was spotted leaving his burial place. He was on his way to torment the locals with loud noises, destroying property and making general havoc, causing great distress. Initially the complainants were laughed at, but when some of the more influential members of the community also began to protest at the unreasonable goings-on, officials started to take notice. Meetings were held and debates conducted, at the end of which it was concluded that nine days should pass from the time of the burial before the body could be disturbed. Masses must be said, but the locals would have to put up with having their behinds pinched and their lamps put out a little longer.

On the tenth day, a large crowd gathered in the chapel. A solemn mass was said to drive the demon out of the body where they imagined it had taken up residence. The body was then examined. Smoke was seen rising form the undead man, though this could have been the smoke from the frankincense that was burned in an attempt to drive off the fearful stink. A butcher bravely stepped forward to open the deceased's stomach and fish about inside in search of the heart. After some time, he had still not found the organ and so was directed upwards through the diaphragm, where he had better luck.

The stench, the supple limbs, the warm body (attested to by the butcher), the dark red blood and the smoke rising from the body, were all taken as clear indications, not of normal decay, but that the deceased had indeed become a *vrykolakas*. They burned the dead man's heart, but to no avail. Things only got worse.

The attacks and the torments continued. People were driven from their homes, more property was destroyed, and no one could get a wink of sleep. Some reflected that they had been wrong to say mass before the heart had been removed from the dead man's body. Clearly, the mass had only driven the evil spirit from his new home temporarily and it had returned to the body afterwards, but that couldn't be helped now. Houses were sprinkled with holy water to try and drive off the undead. Prayers were said and priests fasted. Swords were thrust into the grave in an attempt to pin the body down and prevent his wandering, but nothing helped. Still the attacks continued. One group complained the wrong swords had been used. The handles of the swords looked like crosses and would only keep the devil, who had certainly taken up residence in the peasant's corpse, trapped and unable to flee. Turkish sabres should be used instead. But the residents had had enough. The only remedy was to burn the whole body and afterwards, let the devil remain in the ashes if he could.

What was left of the peasant's much decayed body was taken to the point of the island, known as St George's, and an enormous pyre constructed. The wood was dowsed in pitch and tar in case the body didn't burn fast enough, and they called it the Bonfire of Joy. The locals celebrated as the remains were consumed in the fire and the island was no longer bothered by the *vrykolakas* of Mykonos.

If, after reading the rest of this book, you have decided that you really do have a vampire problem on your hands, this chapter is here to offer some information on how to deal with it in a more efficient manner than the islanders in the story above. As with all things, prevention is better than cure, so we will start by looking at identifying potential problems before your local graveyard becomes infested with the undead.

11.1 FOREWARNED IS FOREARMED

Sometimes, there are indications during life that someone may turn into a vampire after they die. Those who have been attacked by a vampire, even if they survive the attack, may become a vampire when they reach the end of their natural life. Arnold Paole, who we met in Chapter 2, survived a vampire attack and would have lived a long and healthy life had he not fallen from a hay wagon. The earth from the vampire's grave that he had eaten, and the vampire's blood that he had washed his wounds with, only delayed his transformation until after he was dead.

In many traditions, people don't need to wait for a vampire to bite them to be at risk of becoming one of them, some are naturally predisposed to rise from the grave and prey on the living. In some parts of the world, it is a simple accident of birth that can mean some individuals are destined to become vampires after they die. If your birth coincides with certain religious festivals then it is a given that you will rise from your grave and bother the living.

Other potential revenants can be identified at birth because they have teeth, a caul (the amniotic membrane wrapped over the head) or blue eyes. Similarly, red hair was taken as a sign of potential vampirism in the Middle Ages, according to the vampirologist Montague Summers at least. Living a blameless life in the intervening period appears to offer no remedy and becoming an undead blood-sucker is inevitable. Many of these potential vampires are identified as being 'unusual' within what a particular community accepts as 'normal'. All of these things are perfectly natural, but red-haired and blue-eyed people can stand out if the majority of the population have dark hair and/or eyes.

Superstitions easily arise as a way to protect groups of people from potential threats. There is a desire to protect your community from unknowns or outsiders who may pose a threat. Those who don't fit into the accepted pattern can be ostracised as a result. I can see no reason why having the appropriate genetic instructions for red hair or blue eyes would also predispose anyone to becoming one of the undead. Being born with teeth may suggest a greater tendency to bite, but it is a stretch to get from early dental development to walking corpses intent on drinking human blood. Depending on the part of the world in which you were born, having a caul can doom you to a horrible undead existence or bless you with good luck.[†] Arriving into the world on the same day as a revered holy figure or important religious festival is sometimes seen as anything from bad manners to a sign of evil intent. But how poor timing could physically translate into becoming a blood-sucking revenant is perhaps a conundrum for religious philosophy, rather than science.

To add to the list, there were people who had lived a bad life who were also destined to carry on their malevolent existence even after death. For whatever reason, these individuals may have been singled out for undead life, it at least meant that communities could identify those people and take preventative measures. And no one had to worry about anything in particular, unless one of these people died. That was when the trouble would start. However, there were some who might not be so easy

[†]Ireland for luck, and Germany and Austria for vampiredom.

to identify. Some may have tried to hide vampire attacks and there could be any number of other unknown reasons why someone could return from the dead.

Ensuring proper burial rights were observed was often enough to prevent a candidate vampire from becoming a real vampire. If carried out correctly, preparing the body and observing the right rituals around death was seen as the best way to protect the living from the undead. In theory at least, proper burial for everyone protected against every eventuality.

11.2 THE ULTIMATE INSURANCE POLICY

Traditions of how people are cared for after death and the rituals performed for the dead before, during and after they are given over to their final rest, vary across the world. In most parts of the world the dead are buried or cremated. Before this happens, the body is usually prepared in some way, either by washing and/or embalming. In many traditions, where it is family, friends or local experts, rather than professional undertakers, who prepare the body, there are traditions about how it is done, even if there are no specific legal requirements. Many of these practices are rooted in religious beliefs but they also involve practical steps that physically protect the living from any harm that could come from the dead.

Washing the body before burial allows for a detailed inspection that might reveal signs of a certain disease that can alert the community to a new outbreak, or that the death was not due to natural causes. Further measures can then be taken to try and prevent the spread of the disease or start an investigation into a potential murder. Many rituals for preparing the dead include details about how the water that is used to wash the body must be disposed of carefully. It is a very sensible precaution, particularly if death was caused by a contagious disease.

Other practices are not so easy to explain. Many believe that animals, such as cats, must be prevented from jumping over the corpse at all costs, or the body will be back before long in the form of a vampire. In Russia, even birds flying over the body, or in many parts of the world, a shadow cast by a person, can have the same result. Some cultures insist on covering mirrors and other reflective surfaces when there is a death. Shadows and reflections have a long tradition as evidence of a human soul and they mustn't be trapped in a reflective surface after a person has died. Some encourage the soul to leave by opening windows, but in other parts of the world the windows must be kept shut, presumably to prevent anything sinister getting in. In a similar vein, people watch over the dead until it is time for their interment.[‡]

How much time is available to carry out the preparation varies enormously. Some cultures insist on funerals being held within twenty-four hours, perhaps to minimise the time for decay processes to get underway. Some use embalming techniques to preserve the body for a longer period. Chemicals that kill bacteria or restrict fungal growth that would otherwise cause decomposition, are used to allow a greater time between the death and the funeral, but few modern embalming processes are meant to preserve the body beyond interment. Whatever the ritual or process, it usually has the effect of keeping the body in an acceptable, relatively undecayed form, until burial. It also gives a certain uniformity to what is expected of a dead body within a particular culture, perhaps explaining why deviations from these expectations are often taken as signs of vampirism (see Chapter 3).

It may be for these reasons that those who died alone, for example, murder victims and suicides, have been seen as potential vampires. They did not receive the proper attention and rites when they died. If the body is not discovered for some time, decay processes may be advanced and the body could be showing signs that were often taken as indications of vampirism. Those who have died of a contagious disease are also less likely to be attended by the living for fear of contamination. They are less likely to receive the full rites appropriate for the dead and are therefore more likely to return. This may be another way in which vampires have become associated with disease and its spread (see Chapter 8).

Once prepared, there are more rituals associated with the proper disposal of the body. The right words must be said at the right time during the funeral ceremony, but the precautions

[‡]Also an insurance policy in case death has been declared prematurely, or the body shows early signs of being a revenant.

don't necessarily stop at the service itself. In many cases, ensuring that all the necessary funeral procedures are observed is not enough to ensure that a person will not become one of the undead. Further precautions are needed, just to make sure.

Beliefs about when a person has died and when their soul is believed to have departed from the body vary enormously across the globe. Some believe it is at the point that the heart stops beating. In other cultures, it is only when the body has disintegrated that the soul has finally left. A careful watch is often kept on bodies long after they have been buried to ensure the soul is comforted while it remains and doesn't compel the body to try and escape the grave in search of some company. In Romania, spouses will sit by the graves of their recently deceased partner, chatting and keeping them entertained until the soul departs forty days after death. In some parts of Europe, the bodies of the deceased are removed from the grave after a certain period of time – months or years – and further rituals are carried out, such as washing bones in water or wine, essentially a method of checking the body has disintegrated as expected. If you do not have the time to sit around in graveyards and watch over the deceased, just to make sure they really are dead, there are back-up plans available.

11.3 DISORIENTATION

In some traditions, when travelling to the funeral, the body of the deceased must not leave the home through the door, otherwise they will be able to find their way back. Holes are cut in walls to pass the body through, and then repaired afterwards. In some places, mourners are told not to look back at the grave when they leave the funeral or the dead will follow them. All of these methods are thought to disorientate the undead and prevent them from returning to bother loved ones. It is the same reason that, for several centuries, suicides in England were buried at crossroads, to confuse a corpse who might decide to come back and trouble the living. Bodies were also sometimes buried face down for the same reason. Folklore vampires don't appear to be as smart as their fictional descendants.

Sometimes, more extreme measures were needed, and the local geography could help. Running water was thought to provide a physical barrier that vampires could not cross. Why this should be is probably associated with water's purity. For a long time, it was considered an element in its own right, and its clear importance to life and associations with cleansing, both ritual and physical, perhaps gave it a power and association that meant unclean things, like vampires, would be repelled by it. Witches and vampires were generally believed to float. In the case of vampires, this may have been true because the buildup of gases in the body from decomposition (see Chapter 3) would mean any swollen bodies thrown into water were likely to re-emerge and bob up to the surface. In one extreme example, the vampires of Greece were taken to uninhabited islands and left there, trapped on the island by the surrounding water that they could not cross.

But what if you don't want to knock holes in the walls of your house? What if there are no handy islands on which to trap your vampire? Local councils are likely to object to burials at intersections, and in any case the proliferation of signposts that help the living to navigate their way would presumably assist the dead in the same way. Ideally, vampires should be contained in their grave, rather than relying on a poor sense of direction. Measures can be taken to ensure the vampire stays in its coffin and doesn't get the chance to go about its blood-sucking business.

11.4 CONTAINMENT

Before going into the detail, I should stress that digging up dead bodies and manipulating them in any way is breaking several laws that carry severe punishments. What I am describing happened a long time ago and, when people have tried to repeat these kinds of activities in modern times, they have ended up in courts trying to explain what they did and why they should not go to prison as a result. Their arguments are rarely, if ever, listened to sympathetically by judge or jury and a lengthy period of time at her majesty's pleasure awaits.

Many methods of vampire containment involved mutilating the body in some way to prevent it walking, such as cutting the tendons at the backs of the legs, or pricking the soles of the feet. Pricking the body with needles, pins, nails through the forehead or stakes through the chest were believed to stop dead relatives transforming into undead relatives. These methods would in most cases help to prevent the body from swelling by releasing the gases built up from the frenzy of bacterial feeding going on inside the body, a classic sign of vampirism in folklore (see Chapter 3).

Stabbing or staking vampires was not always about 'killing' them, it was to pierce the skin and prevent their transformation or to pin the body in place and stop their wandering. As we have seen, just because a body is dead, this does not mean it will stop moving, though these movements don't amount to getting up and wandering about. Fixing the body to the spot isn't what stops it from walking about, but this may have reassured terrified locals. Archaeological evidence shows this certainly happened in the past. In England, up until the nineteenth century, the bodies of suicides were buried at crossroads with a stake through the heart. Romani in Yugoslavia, who provided something like a vampire elimination service, would place the tip of their stake over the centre of the grave and push through the overlying soil in the direction of the body below. Their method aimed for the stomach, to pin the vampire, while preventing splashes of the vampire blood that may contaminate them.

Some traditions demand that, rather than avoiding the vampire blood, you should actively seek it out in order to properly protect yourself, as Arnold Paole did in 1732 in Serbia. He ate earth from the grave of the vampire that attacked him and washed his wounds in the vampire's blood. This protected him as long as he was alive,[§] like some kind of vampire-vaccine, though that is not exactly how vaccines work. The idea of a vaccine is to prepare the body against a potential biological threat by introducing a small amount, or a weakened form, ahead of time. This prepares the immune system should the real, fully virulent form of the virus, come along later. What it doesn't do is delay infection or the effects of an infection. Vaccines are of no use once the body has already been infected.

If the idea of drinking blood direct from the body in the grave does not appeal, then it can always be cooked. In parts of Poland,

[§]It would seem that drinking vampire blood can either turn you into a vampire, or protect you from its predations, depending on which part of the world the blood-sucker is buried in and to which vampire tradition they belong to.

vampire blood was used to make a bread that was eaten to prevent further attacks. This could be taken as an analogy of attenuating or inactivating a virus to make it safe for use as a vaccine. By weakening a virus or changing it so much it can no longer cause an infection, the molecular makeup of the virus is introduced into the body safely but the immune system can still examine it and remember it.

If you want to avoid direct contact with the body, some believe that shooting bullets into the grave will work just as well as stakes or nails; a few claimed that all that was necessary was to shoot over the grave. Silver bullets may or may not help as, though links between vampires and werewolves exist in the folklore, it is not clear if they are truly the same thing and can be stopped in the same way.¶

Mutilating a body is looked on particularly unfavourably by the general public as well as the authorities. If you are at all squeamish at the idea of cutting into or otherwise maiming a dead body, then there are less invasive methods that can be used to contain the vampire, but, be warned, opening the grave to carry them out is still illegal.

A knife, sickle, stake, or whatever sharp or pointed instrument you have handy, can be placed over the corpse without actually touching it. With careful placement, if the undead decide to get up one evening, they will be impaled, pierced or cut preventing their exit. Certain plants have been associated with repelling vampires and can be thrown into the grave or scattered around the home. Many of these plants are known for their thorny stems – dogrose, blackthorn, whitethorn, wild rose – which would tangle in the vampire's shroud and leave them struggling to get free. Hawthorn stakes could also be made and driven into the soil above the grave – just in case. The *penanggalan* of Malaysia, a flying head with entrails strung out behind it, can be caught using thorns placed on window sills that catch on the entrails and hold it in place before it can feed on children's blood. If it is the wrong season for these plants, or they simply don't grow in your neighbourhood, the body can be weighted down with stones, tied up or pinned down with arrows. Evidence of all these practices has been found in graves by archaeologists.

¶There will be more about silver and bullets in Chapter 12.

Maybe you have already eaten and the idea of blood bread simply can't be stomached. Or, you fear that methods of entangling or trapping the vampire may not be secure enough and it could later emerge Houdini-like to return to their usual activities. You may even have some sympathy for the vampire and you would prefer to use a method that did them no harm, but also protected you and yours from harm. There are plenty more vampire-friendly options available.

11.5 FATAL DISTRACTION

Many folklore vampires have been thought to show evidence of chewing, which was taken as an indication of their intent to chew others. In actual fact, it was perfectly normal movement of the body, perhaps combined with leakage of decomposing material from within the body, or other animals chewing on fingers and other extremities. But, if there is any doubt or worry that vampires will turn their attention to living people, one idea is to give them something else to chew on. Bricks, stones, coins and potsherds can be placed in the mouth. Clods of earth or books placed under the chin to close the mouth can stop the undead from chewing their shroud, themselves, their neighbouring corpses or living people in the local area. The vampire would not be 'killed', but it would continue its undead existence being very, very hungry. Apparently removing these bricks, coins or other impediments was beyond the folklore vampires that were treated this way. The bodies that have been uncovered centuries later by archaeologists would suggest that many of them did indeed simply put up with foreign objects being shoved in their mouths (see Chapter 3).

Some vampires have an obsession with neatness and order that can be exploited to keep them occupied in their grave and away from the living. Having identified the grave of the vampire, and perhaps not wanting to get too close to the body for fear of being attacked, all you need to do is give them something to occupy their time. Fishing nets left in the grave will help entangle the vampire, but they might also be compelled to untie all the knots. Hanging the nets over doorways or windows will distract the vampire if they have managed to get out of the grave and track you to your home. But be careful, not every vampire is interested in the same thing. Knot-obsessed vampires seem to be found mostly amongst Romani communities, and a few in northern Germany.

In eastern Europe, vampire hunters would steal a single sock from the vampire's body. The sock would be weighted down with stones or soil from the grave and thrown outside the village boundary, preferably into running water. The vampire, infuriated by his uneven footwear, would spend all his waking hours searching for the missing sock rather than sucking blood.

Other vampires suffer from arithmomania, if the children's programme *Sesame Street* is anything to go by. *Sesame Street* is a highly respected and very educational programme in many ways. We can all learn a lot from the programme, including how to protect yourself from vampires. Everyone knows that Count von Count, the Count in Sesame Street, loves to count. If the Muppet in question was ever to think of sinking his felt fangs into anyone's neck, simply giving him lots of things to count would keep him distracted while you escape. The creators of this iconic kids' programme have certainly done their homework. Many vampires are indeed obsessed with counting.

All you need to do is pour lots of small objects into the coffin for the undead occupant to count. The type of small object doesn't seem to matter very much. Some have used poppy seeds, possibly because of their association with narcotics, that would hopefully help the vampire to rest. Poppy plants, including the seeds, contain opiates, including morphine and codeine, which have sedative properties alongside their well-known pain relieving abilities.‖ There are, however, few indications in the history of vampires that suggest the vampire was expected to eat the seeds. You also have to eat an awful lot of seeds to feel the effects of the morphine within them.** If you can't get hold of poppy seeds, you can use other seeds, small stones, sand, rice or whatever is available. Just as with the Count in Sesame Street, as long as there is something to count, the vampire will be compelled to count it.

Arithmomania afflicts vampires from many different cultures all over the world. The *soucouyant*, the vampire-like creature

‖Morphine was named after Morpheus, the Greek god of sleep and dreams.
**About 75 g worth, I have been told.

from the Caribbean that we met in Chapter 7, can similarly be kept safely occupied by seeds and rice scattered around the homestead. In South America, the *asema* is a vampire-like creature that travels about as a bright ball of blue light. Small stones or seeds can be scattered outside the home and the *asema* will be compelled to tidy them into small piles. It's an easy way to protect yourself from attacks from revenants, and you don't even need to know which grave is housing the troublesome bloodsucker. The key is quantity. Some say the undead count at a rate of just one seed or grain a year, but others are much quicker, so use lots. Provided there are enough objects to count or tidy, the vampire will be occupied until daybreak and potential victims can rest easy in their beds.

Many of the above methods rely on identifying exactly who the vampire is. Finding the culprit is key, but that can be easier said than done. There is more to tracking down vampires than simply hanging around graveyards late at night and waiting for them to appear.

11.6 VAMPIRE TRACKING

Evidence from cases reported in the eighteenth century, and the vampire fiction of the nineteenth century, would suggest that usually the first anyone knows about a vampire outbreak is when people start to become sick. Sometimes, the culprit was easily found because they were simply the first person to have died during the recent outbreak. In the eighteenth century, the vampire was usually known to the locals. In small communities it would be common knowledge who had died, and of what cause. Individual vampires might be spotted prowling about the streets late at night and be recognised as one of the recently deceased. As if to remove any doubt, victims often named the person who was creeping into their bedrooms to feed on them. It was then a simple case of going to the graveyard and checking the suspected corpse for signs of vampirism.

As we have seen in previous chapters, this examination was usually carried out at night, when any suspected vampire might be expected to be out and about preying on their next victim. Surely, if these former members of the community really had become revenants, their coffins should have been empty, as it was in *Dracula* when the team led by Van Helsing go to the tomb of the recently vampirised Lucy Westenra. But in folklore accounts of vampires, they are always found to be at home when the vampire hunters call.

Vampires, outside of fiction and film, were never caught in the act. It might seem obvious to wait for the vampire at the place where an attack was expected. Even if the vampire escaped, their victim would have at least had a respite for one night. But folklore vampires don't seem to have gone to much trouble about planning their attacks or avoiding capture. Having identified the vampire, the grave was easy to locate and the undead were always waiting there placidly in their coffins until the stake went through their heart and they offered a brief moment of protest. It seems very considerate on the part of folklore vampires to make everything so easy for the vampire hunters.

Greek vampires were even better organised, working to a strict timetable. Vampire hunters in Greece didn't need to bother with late-night trips to graveyards on any day except Saturday, when they knew the vampire would be at home. Apparently this breed of vampire didn't need sleep the other six days and nights of the week, and the local populace simply had to put up with the vampire's antics until the weekend when the undead were compelled to rest in their coffins.

But knowing when a vampire was home didn't help with knowing which home they were in. There could still be a lot of wasted effort digging up the recently deceased, who had nothing to do with the vampire attacks and were quietly resting in peace. To save the effort of a lot of late-night spade-work, there were other ways to identify which grave may be home to the undead.

When hunting for the *strigoii* in Romania, a horse was led around the graveyard. This horse had to be black. If you were in Albania and employed in the same task, the horse had to be white. Traditions varied across Europe, but in any case the horse had to be of only one colour, and it had to be a mare that had never been mounted, or a young horse that had never stumbled. In some cases, a virginal youth also had to be riding the horse. The horse would move among the gravestones freely until it reached the grave of the vampire and then it would refuse to cross the grave no matter how much it was coaxed, lured or cajoled. This may have been a perfectly sensible reaction on the part of the horse. The soil in a recently made grave would not be as compacted as the ground around it or the soil on older graves. A horse, unsure of its footing may well avoid walking on ground that looks uncertain. Or perhaps the stench of a decaying corpse emanating through the soil of a relatively shallow grave, may have halted it in its tracks. A horse might also be more sensitive to the sound of the corpse groaning or chewing from under the soil.

If the trouble of recruiting unblemished horses and youths sounds like too much trouble, then a good torch may be all you need. Several reports from the eighteenth century stated that a vampire's grave could be identified by the appearance of small holes in the soil above the coffin. These holes, about the width of a man's finger, were said to be the revenant's means of exit and entrance to their coffin. This is perhaps where the shapeshifting abilities of blood-suckers came from (see Chapter 7) in order to explain how the vampire was able to leave its grave without disturbing the soil. There is also a more rational explanation for the holes than vampires turned into fog or stretched out like sinister spaghetti. The holes could appear naturally as the freshly dug soil, recently replaced in the grave, settles over the coffin. The soil may be even more disturbed than just a few holes as sudden changes in weather, such as heavy rainfall or a dry spell, can alter the way the soil gets moved about without the aid of the undead.

Sometimes, all the horses and all the men searching for disturbed soil still couldn't track down the suspected vampire and mass exhumations of all the recently interred were ordered. Considering vampirism could be passed on to new victims, it was a sensible if time-consuming method of eliminating all the vampire contagion in the local area. For example, the case of Arnold Paole, where some cows and sheep were overlooked during an initial investigation and vampire cleansing, resulted in further outbreaks a few years later. It shows you can never be too careful. Peter Plogojowitz's reign of terror was brought to a swift end when he and all of his victims were disinterred. It was not enough just to destroy his remains, his victims also had to be carefully dealt with.

Tracking a vampire to its grave after a spate of attacks can quite literally involve shutting doors after horses have bolted.

No one wants to be the first victim that alerts everyone else to the unwelcome return of the undead. How do you make sure you aren't the one who wakes up one night to see a pale face peering in at your window and licking its lips?

11.7 PERSONAL PROTECTION

Vampires are surprisingly polite. You might not expect predatory blood-suckers to respect personal boundaries, but many vampires have to be invited to cross a threshold. The obvious way to prevent an unwelcome visit from the undead is to not ask them in if they come knocking. It may seem a straightforward safeguard, but vampires are tricky, and even tacit invites are acceptable, and once given they cannot be taken back. If the vampire crosses your doorstep or windowsill just once, they will have unfettered access to your home from then on.

Things were a little different in the Caribbean. There, the vampire-like *soucouyant* could be deliberately invited into the home to trap it there. The trick was to invite them into your house during the day – an important detail – and while they were inside, to turn a cocoyea broom (a traditional broom made from coconut leaves) upside down. The *soucouyant* would be trapped inside the house and unable to transform themselves until the broom was turned the right way and they could escape. But be careful that you have the right variety of vampire before opening the door to them. You don't want to be trapped inside a house with the non-transforming, permanently-out-for-your-blood version. As the film says, *Let the Right One In.*[††] If you have made a mistake, then your only hope is to keep the vampire away from your person.

Religious symbols, particularly crosses and crucifixes are thought to repel vampires (see Figure 11.1). This might have been more effective in the past when belief in a god was near universal. Your church, temple or mosque would have been a symbol of sanctity and protection. Whatever vampires might be in your part of the world they were never on the side of the good guys. It was therefore only reasonable that religious symbols that were good and holy would repel them. But, presumably, you had

[††]An excellent Swedish vampire film. Don't bother with the American remake.

Figure 11.1 Dracula (Bela Lugosi) reacts to a crucifix held by Professor Van Helsing (Edward Van Sloan). A lobby card for the 1931 film *Dracula*.

to use the right religious symbol for the right revenant. Muslim vampires may be unaffected by water blessed by a Christian priest and atheist vampires may be indifferent to the power of the Bible, for example. Nevertheless, symbols from someone's faith may have had a very comforting effect for the potential victim, even if they made no impression on the undead.

Another way to repel these revenants was to make yourself as personally unappealing as possible. In Romania, drinking wine, and the cheaper the better, was thought to taint the blood and make it unpalatable to the undead. In Pomerania, blood brandy, literally brandy with added blood taken from the vampire by soaking its shroud in the liquid, had similar protective properties. Whether it was the blood or the alcohol that was putting the vampire off its dinner is unclear. The high alcohol content of the brandy would at least go some way to protect the living person who drank this concoction, as it would help kill some of the bacteria that would inevitably soak into the shroud along with the blood. If blood brandy and cheap wine doesn't appeal to your palate, other traditions have tastier anti-vampire snacks. Potential victims of the South American *asema* can eat certain herbs that will taint their blood and make it unappetizing for the vampire. And, of course, there is one commonly eaten herb that is particularly well known for its vampire-repelling properties.

11.8 GARLIC

This plant has been singled out amongst the many herbs and spices that humans have put to use because of its pungent taste and smell. Quite apart from flavouring food, garlic has found many diverse applications over the centuries. Garlic plants are members of the *Allium* genus, named from the Greek αλεω, meaning 'to avoid'. For thousands of years, garlic has been used in a whole host of guises to repel poisons, disease, insects, people and evil spirits.

Mummies have been prepared using garlic as part of the embalming process. Two thousand years ago, the Chinese were threading garlic bulbs onto red string and hanging the cords in doorways as insect repellent. In other places it was used to attract certain insects. The Romans ate garlic for its supposed aphrodisiac properties, though Pliny the Elder, the Roman author and natural philosopher, wrote that it was also eaten to discourage unwanted attentions. Pliny also claimed garlic could keep off scorpions and serpents and cure madness. Aristophanes, the Greek playwright of the fourth century BC, joked about garlic plasters being used to treat boils on the skin. Greeks ancient and modern hung garlic in doorways and draped the bulbs round newborn babies' necks to ward off the evil eye. And, of course, it has been smeared on doors and windows, draped round necks and thrown into coffins to keep vampires at bay. Can one plant really do so much?

Temporarily setting aside garlic's potential as a prophylactic against the supernatural, there might be more to some of these traditional remedies than meets the eye, or nose. The strong smells and taste associated with garlic are down to the high number of sulphur compounds it produces as it grows. Sulphur-containing molecules are amongst some of the smelliest known. Many animals are particularly sensitive to sulphur compounds because they are often produced as a byproduct of bacterial decomposition. Sensing these compounds is a warning that what you were about to eat might not be too fresh. Some insects will be repelled by such scents, but others will be attracted.

Garlic plants produce these sulphur-based molecules as secondary metabolites – compounds that are not needed directly for the plant's growth or health, but have a secondary role in repelling animals that might eat it, while still attracting pollinators. Garlic really does have the potential to repel insects you don't want in your house, while attracting some you might want to pollinate your crops.

Of the many sulphur-based molecules that can be found in the humble garlic plant, there are some that have biological activity in humans. Garlic is not a universal antidote to poisons, but it can allay the symptoms of cyanide and arsenic poisoning because of its high sulphur content. Adding a sulphur atom to cyanide hugely reduces its toxicity and arsenic's strong affinity for sulphur can trap it before it can have its devastating affect on the human body. Eating more garlic in regions where the water supply is commonly contaminated with arsenic, such as around the Ganges delta, may help allay the problems of chronic arsenic poisoning. And specific sulphur compounds can have other biological benefits.

This is perhaps why garlic is the most commonly used plant in the history of folklore medicine and has been part of numerous remedies and treatments from the time of the ancient Egyptians. It is mentioned in 22 remedies written in the oldest surviving medical text, the Ebers Papyrus dating from around 1550 BC. Garlic pills are still among the top-selling herbal supplements today. While garlic has many compounds that have potential therapeutic benefits, the medical claims made about garlic-based herbal treatments are often exaggerated or just plain wrong.

The distinctive smell produced when cutting garlic is due to a compound called allicin. In the plant, an allicin precursor and an enzyme alliinase are found in equal amounts but they are kept separate. When cells are broken, by cutting or crushing the plant, the precursor and enzyme can mix, and the result is allicin. This compound has been shown to have significant antifungal and antibacterial activity. Perhaps this is why garlic was thrown into the graves and stuffed into the mouths of suspected vampires. Such properties could have hindered the rapid growth of bacteria or fungi that occurs after death and can lead to the bloat stage of the decay process so often associated with vampirism. However, the antibacterial activity may be significant but it is only 1% of the activity of penicillin, although unlike penicillin, it is just as effective against Gram-positive and Gram-negative bacteria.[‡‡] And allicin is still considerably less effective than other antibiotics, not just penicillin.

Before the advent of antibiotics, garlic therapies were sometimes used to treat the bacterial infection tuberculosis, a disease often associated with vampire outbreaks (see Chapter 8). Inhaler masks were given to patients that contained a sponge soaked in garlic juice. Good nursing was a much-preferred course of therapy, by everyone I would imagine but especially the tuberculosis patient, and was considerably more effective.

Allicin is not the only compound that could have been of benefit, and not just to tuberculosis patients. Another chemical in garlic, ajoene, has shown anticancer and antifungal properties. In relation to a vampire's favourite substance, blood, diallyl disulfide and trisulfide can be converted by the red blood cells into hydrogen sulfide. Hydrogen sulfide is very toxic, but in suitably small quantities, and in the right places, it plays a vital role in signalling between cells. It can cause blood vessels to dilate and improve blood flow. Hydrogen sulfide from garlic extracts has also been shown, in mice at least, to prevent some of the damage caused by heart attacks.

Before you rush to the supermarket or garden centre to get hold of a miracle home remedy, however, there are some things about garlic compounds you should know. The activity of garlic compounds can have good and bad effects. For example, garlic extracts have been shown to inhibit the growth of *Mycobacterium avium*, which can cause life-threatening infections in AIDS patients.[§§] Garlic supplements have also been shown to inhibit the activity of saquinavir, an anti-HIV protease inhibitor drug, and possibly to interfere with other sulphur-based drugs. A review of medicinal uses of garlic and its compounds in 2007 showed that the benefits are slight and not clinically significant. Garlic may not just be ineffective, it can also be damaging.

[‡‡]Bacteria can be classified into two broad categories based on whether they give a positive reaction to the Gram stain test. The differences are in the cell wall and can affect how well different antibiotics interact with the bacteria.

[§§]Sales of garlic supplements soared during the 1990s AIDS epidemic.

Many will know from first-hand experience that getting raw garlic juice in a cut will sting. When it comes to the undead, perhaps these severe irritant properties were seen as a powerful way to repel them. Allicin may have some anti-bacterial activity but it is also a potent skin irritant. A number of medical reports have been written after patients have arrived in hospital with severe burns as a result of eating or applying too much raw garlic.[¶¶] Care should be taken if raw garlic is being applied to the living, even if it is as a last line of defence against vampires.

When it comes to the undead, whether bulbs are thrown into coffins, smeared around windows or door frames, hung round the neck, flowers placed in vases or the cloves cooked in meals, garlic is a fairly easy, if unpleasant, prevention measure. No one minds a bit of extra garlic in their dinner, but the idea of spending night after night surrounded by the suffocating smell may feel like a bit much. Given vampires' longevity and obsessive nature, they can presumably wait for the night that you forget and let down your garlic guard. If you can't face a lifetime saturated in this powerful perfume, and all other preventative measures have failed, there is only one course of action left – the vampire must be killed. Just how you can kill something that is already dead is the subject of the next chapter.

[¶¶]Just one or two cloves of raw garlic can irritate the lining of the digestive tract.

CHAPTER 12

Slaying

'... and there is no way, it seems, to get rid of these troublesome visitants, but by digging them out of their graves, impaling them, cutting off their heads, taking out their hearts, and burning their bodies.'

Reverend Father Augustin Calmet, *Dissertaions Upon the Apparitions of Angels, Demons and Ghosts, and Concerning the Vampires of Hungary, Bohemia, Moravia and Silesia*

If there is something strange in your neighbourhood, who do you call? Well, in Bulgaria, you would call a sorcerer. These highly trained individuals went to elaborate lengths and endured great personal risk to trap and destroy the blood-sucking undead. Eliminating Bulgarian vampires was the work of professionals. Amateurs should stay at home eating garlic.

If the people in a Bulgarian village or town realised they had a vampire problem, they would quickly retire from the field and leave all the hard work to the sorcerer, and it was indeed hard work. When the vampire in question had been identified, the sorcerer would make their preparations. A bottle, a religious icon and some of the vampire's favourite food were collected. The sorcerer would then lie in wait, hidden away near where the vampire was expected to appear on its nightly forays, and keep a careful watch for when the troublesome revenant would wander past.

Vampirology: The Science of Horror's Most Famous Fiend
By Kathryn Harkup

Published by the Royal Society of Chemistry, www.rsc.org

At the sight of the vampire, the sorcerer would spring from his hiding place brandishing the icon in one hand and the bottle in the other. In fear of the religious symbol, the vampire would flee and the sorcerer would chase behind, along the streets, through houses and up trees if need be. When the vampire was eventually cornered, the sorcerer would offer it two choices: the icon or the bottle containing some nice fresh blood. The vampire obviously chose the bottle. Genie-like, it would squeeze itself inside and feed on the morsel of blood. The sorcerer would then stopper his exit with a cork, the underside of which had been carefully covered with a fragment of the icon. The vampire-filled bottle was then thrown into a fire to destroy it.

Some vampire hunters had it easier than others. Depending on what part of the world you were in and what type of vampire you were dealing with, sometimes the destruction of the 'head' or 'king' vampire was enough to drive out the vampiric taint from their victims too. The remaining vampires in the clan or coven would be returned to being dead, rather than undead, after the destruction of their master. Other vampire hunters, however, had to track down and destroy every last vampire and victim to rid the area of the epidemic (that is, whether the victim be human, sheep or cow, see Arnold Paole in Chapter 2).

As we have seen, there are many varieties of vampire and there are just as many methods of dispatching them. Vampire hunters need a specialised set of equipment and expert knowledge to deal with the vampire in question in an appropriate way. Fortunately, Bulgaria did not have the monopoly on vampire slayers. Other countries relied on other specialists within their community. For many of the nameless individuals who have worked to eradicate the undead, it was just a job. A very skilled job fraught with danger, but still a job. Even slayers have bills to pay. A few, however, like Buffy Summers, Professor Van Helsing and Blade, have stood out amongst their fellows.

Many of these slayers, like the vampires they hunt, were born with specialist powers that help them track down and destroy the undead. Others have dedicated themselves to a lifetime of study and intensive training to achieve the same feats. It may sound glamorous and exciting, but it involves a lot of hard work and considerable personal risk.

Before taking on the role of slayer, please take a little time to consider your actions. Destroying vampires might be a noble calling, but mistakes can be made. You have to be absolutely certain that the person you are about to maim or kill really is a vampire. Many of the methods discussed below will harm the living just as effectively as the undead. Mistake an actor in a vampire role, or someone just enjoying the vampire aesthetic, and you could be left with a dead body on your hands and an awful lot of explaining to do.

But if you cannot be dissuaded, what knowledge, skills and weapons might you need? How do you kill something that is already dead? And just who might be best suited to such a role?

12.1 BORN TO SLAY

As a kind of natural counterbalance to the threat of vampires, many communities also produced the means of defeating them. It was no coincidence that the small American town of Sunnydale produced a slayer among its inhabitants, situated as it was above the mouth of hell and positively overrun with vampires and other demons.

Europe also had a notable vampire problem, particularly in the eighteenth century, and produced its own slayers. *Dhampirs* were the children born to one human and one vampire parent. They inherited some of the beneficial powers from their vampire parent but none of the bad habits that make them such a menace. The specifics of their abilities and limitations varied from region to region. In general, they were mortal but had powers that could be used to detect and eliminate vampires. They could see invisible vampires, knew special ceremonies that could trap and destroy them and were immune to their predations. They could therefore personally fight a vampire without running the risk of being destroyed or converted into one. *Dhampirs* would travel around offering their services to rid communities of vampire infestations. After many eccentric rituals, quite a bit of running around and giving the impression of wrestling with an invisible foe, the *dhampir* would declare the vampire destroyed and collect a huge fee.[†]

[†]The last *dhampir* ritual was apparently carried out in 1959 in Kosovo.

The most famous *dhampir* is Blade, a rather more sophisticated slayer than his eighteenth-century predecessors. The fictional character began life in Marvel comics in 1973, and went on to star in three films – where he was played by Wesley Snipes – and a TV series, where the role was taken on by Sticky Fingaz. Blade's backstory relates how his mother was attacked by a vampire during her pregnancy and several vampire characteristics were passed on to him at birth. Blade was mortal but had a much-extended life, he enjoyed increased strength and complete resistance to vampire attacks, as well as many of the substances that vampires feared such as sunlight, garlic and silver. He then augmented his natural skills with further training in a range of martial arts. Over the course of his career through comics, films and TV series, he used perhaps every type of weapon, both designed and improvised, that have been known to kill the undead. However, he made particular use of specialist weapons – which were mostly bladed (hence his name) – to kill his vampire foe.

In the film series, science was able to add to Blade's stock of weapons when haematologist Karen Jenson, who collaborates with Blade, discovered that mixing a solution of the chemical ethylenediaminetetraacetic acid (EDTA) with vampire blood caused it to explode. Syringes full of bright blue EDTA were deployed by Blade to defeat the vampires in spectacular fashion. EDTA would appear to offer a clinical, relatively simple, but very messy, way of destroying a vampire. Could science have found a new way of destroying the undead?

12.2 CHEMICAL KILLERS

EDTA, or ethylenediaminetetraacetic acid, is a long name for a relatively simple molecule. It is a colourless solid that dissolves easily in water to form a clear solution (not blue as in the film, but a bit of artistic license is understandable to make it look a bit more interesting). EDTA might not look like much, but it is very useful. It is a common additive to shampoo and cleansers; it is used in medicine to treat mercury and lead poisoning, as well as in a complex with radioactive chromium to monitor kidney function; it has also found roles in the textile and cement industries. The reason there are so many applications for EDTA is because of its ability to grab hold of unwanted metal ions, and that is all down to its structure. It has six atoms – two of nitrogen and four of oxygen – positioned on a carbon framework in such a way that it readily wraps around any stray heavy metals, like a hand grabbing hold of a baseball. And once bonded to the heavy metal, it holds on tight.

This is how EDTA can be used to remove toxic lead, mercury or chromium ions from anyone who has been poisoned by them. It strips out the dangerous ions and allows them to be safely transported through the body to the kidneys, where they can be excreted. EDTA will also grab hold of excess iron in blood, which is why it is used to treat people who have a buildup of iron in their body through receiving many blood transfusions. Perhaps this is where the link to vampires came in. They have certainly been on the receiving end of a lot of blood, and are very likely to have more iron in their body than the average human. If vampires have a particular biological need for more iron, then depleting it, or at least removing it from general use by wrapping it up in a molecule of EDTA, could disrupt a vampire's normal metabolism. In this respect, it could be considered a poison that is more specific to vampires than humans. It seems unlikely that it would cause a vampire's blood to bubble and swell and eventually explode, but it does make for a good visual effect in the films.

EDTA might not kill vampires, but it is good to know that your bottle of shampoo or shower gel might contain an ingredient that would at least make them feel a little unwell.[‡] Before Jenson's scientific help gave Blade a new weapon, he had been getting by with more traditional methods of dispatching the blood-sucking undead, including another chemical method: silver.

12.3 SILVER

Blade may have been inspired by earlier folklore vampire traditions that used silver, particularly silver bullets, to destroy vampires. Silver bullets are the traditional way of dispatching a

[‡]Please don't inject anyone with shampoo or any other EDTA-containing substance. It might not kill them but it is likely to make them very unwell.

werewolf but, as we have seen, vampires and werewolves are often found together in films and folklore stories of beings that prey on the living. It is unsurprising that some of the qualities of both these creatures also coincide. The TV series *True Blood* made use of the association and had their vampires burned and paralysed by any silver simply touching their skin.

Silver is used in medical devices because it can protect people from bacterial infection. Nanoparticles of silver can be impregnated into the plastic that is used to make catheters, or the tracheal tubes used to help people breathe, as well as other medical equipment. The silver kills bacteria by inhibiting important enzymes that help it produce nutrients, and can also bond to the bacteria's genetic material. It reduces the chances of infection in hospital environments and can even be added to cleaning cloths and socks to keep them smelling fresh for longer. As a salt formed with chlorine or bromine, silver was also the photo-reactive material used in photographic plates and film in the 1920s and 1930s. The antimicrobial properties helped to preserve Bela Lugosi's image as Dracula and gave him and his cinematic vampire immortality of a sort.

In terms of silver as an anti-vampire remedy, the antibacterial properties may go a small way towards inhibiting the all important decay processes that produce the changes that were often used to identify vampires in the eighteenth century (see Chapter 3). But the damage done to bacteria can only be a minor factor when compared with the damage caused by a bullet, of whatever material, being fired into a body. The puncturing of the body would have prevented the buildup of gases, as they could easily escape through the bullet wound and the body would be less likely to swell into the characteristic inflated shape of folklore vampires.

If vampirism is the side effect of an infection, silver may cause damage to the vampire's body by killing off the source of infection and transformation. Perhaps silver objects held against the skin, as seen in *True Blood*, could cause some localised damage as a result of destroying the bacteria. The ability of silver objects to completely immobilise a vampire may be a placebo effect. For example, people who have been told they have touched poison ivy can develop skin blisters even though they have not come into contact with anything that could cause skin irritation. Belief in the person telling you the information, and knowledge of what should happen if you touch poison ivy for real, can produce a real physical effect in the body. The vampire believes silver will affect them, and so it does.

The extreme reaction to silver seen in *True Blood* is not common for folklore or screen vampires. For most, including Blade, the slaying ability of silver is confined to its use in bullets. In this regard, it would seem to offer no particular advantage over ordinary lead, copper or steel bullets that would also be much cheaper and easier to get hold of.

Of course, one of the most common chemical methods of killing a vampire is holy water. There is a long tradition of holy water being used to deter and even destroy the undead. Depending on the film or folklore tale, holy water can cause a vampire to smoke, catch fire or dissolve completely. Blade seems to have made no use of this potentially powerful weapon. What if he needn't have bothered adding EDTA to water, or buying expensive silver bullets. Could he have destroyed vampires using what comes out of the tap and an obliging priest?

12.4 WATER, WATER EVERYWHERE

As any chemist will tell you, water is special stuff. For thousands of years, it was considered an element in its own right, but the discovery that each water molecule is made of two hydrogen atoms and one of oxygen did not diminish its importance. Water has a range of physical and chemical properties that gives it a unique status. Standard measures of volume, temperature, energy and weight have all been based around water. For biology, water is perhaps even more important than oxygen.

The purity of water, its ability to cleanse and its essential role in life means there are long-standing and complex traditions associated with it in folklore. In the past, it has been believed that this clear, pure substance would not tolerate impurities. Unclean things would be rejected by water – women that floated were witches, bodies that floated were vampires. Out of all this developed the idea that vampires could be stopped in their tracks by running water. Just sprinkling holy water on a vampire's grave or coffin means it is completely sealed off from them and they can no longer return there. And, holy water has been attributed with much greater anti-vampiric properties. It is lethal to them.

Water is actually quite a reactive molecule. It dissolves, corrodes and reacts with a fantastic number of other substances, it's just that, because it is so common, we are used to this behaviour. But, this high reactivity can cause problems for some people. There are people who have a particular sensitivity to water and have extreme reactions to it when they shower or swim. For those who suffer from a condition called aquagenic urticaria, water brings the skin out in hives. Another form of the condition, known as aquadynia, causes widespread pain that can last for 15 to 45 minutes after contact with water.

Though the mechanism is unclear, it has been suggested that a chemical reaction between water and sebum (the oily substance secreted by the skin) creates a product that triggers an allergic reaction by the body. Other theories suggest it is changes in the osmotic pressure around hair follicles that provokes a reaction; or that water-soluble antigens in the epidermis are transported across the dermis dissolved in the water, causing a release of histamines (the cause of symptoms in allergic reactions, hence antihistamines can help alleviate them).

Though the symptoms described in aquagenic urticaria can be severe and very unpleasant, the reaction some vampires have to holy water is much more extreme. The condition in humans is thought to have a genetic component. Perhaps vampires, if they are a distinct species, or subspecies of human (see Chapter 9), could have an extreme variation. It would also explain why some vampires seem indifferent to holy water, while others melt like the Wicked Witch of the West. Could a similar, but more aggressive, reaction occur between the skin secretions of some blood-suckers and holy water?

Holy water may be very different in terms of its spiritual power, but blessing would make no difference to the chemical or physical properties of the water. The different degrees of reactivity must be due to different skin secretions. Perhaps a change in vampire physiology means that, rather than secreting sebum, or in addition to it, some other substance is found in their skin that provokes an even stronger reaction with water.

There are certainly substances that react vigorously with water. Many people will remember the reaction between water and sodium metal from their school days. A small lump of this silvery metal dropped into a dish of water will spark, smoke and soon catch light. Vampires are unlikely to be sweating sodium metal out of their pores, but there may be some compound that causes the smoking and extreme damage seen in films like *The Lost Boys*. This compound, whatever it might be, could perhaps be produced in fatty tissue, away from the huge amounts of water normally found within the body.

With enough water, like the bath-full used in *The Lost Boys*, and enough secretions, it might be possible to cause sufficient damage to destroy a vampire. The film has one vampire completely dissolved in the bath so the remains can be disposed of down the plug hole. It is certainly possible to dissolve a human-sized body, several serial killers have done so, but they usually used strong acids that could disintegrate all types of material found in the body, from skin to bone. You would also need a lot of acid to dissolve something as big as a human body. The reaction between water and a vampire's skin secretions may create a powerfully corrosive product, and you may have plenty of water at your disposal, but it isn't the water itself doing the dissolving. It is the product of the reaction between water and the secretions that does the damage. It's doubtful a vampire could secrete enough to dissolve its own body.

Holy water seems an uncertain weapon when it comes to fighting vampires, but it does have one very important advantage. If you have made a mistake and the being you are attacking is not a vampire, the worst outcome is some very wet clothes. Most slayers rely on more certain methods of dispatching their foe but it comes with much greater consequences if any identities are mistaken.

All the big-name slayers are very different in terms of nationality, age and background, but they all have one thing in common: they all use stakes. In Blade's early days, he impaled his opponents on teakwood daggers, mahogany stakes or, in an emergency, broken broom handles. Staking seems to be an almost universal way to dispatch, or at least restrain vampires, but there is more to it than meets the eye. Both the technique and the tool are important.

12.5 YOU CAN STAKE YOUR LIFE ON IT

Staking is perhaps the most obvious and common method of destroying a vampire. The way some view it, however, staking is not the end of the vampire, it is merely pinned in place and unable to get hold of rejuvenating blood. As a consequence, decay processes take over and the body slowly crumbles to dust. In other cases, the vampire continues its eternal undead existence indefinitely but is prevented from roaming around and bothering the living. Should the stake be removed at a later date, there is a chance the vampire could return to terrorise the living once again. Mostly though, the aim is to destroy the vampire.

The target of the stake is usually, but not always, the heart – the organ most closely associated with life and the end of it. Until the invention of effective cardiopulmonary resuscitation (CPR) techniques, defibrillators, pacemakers and suchlike, that could keep the heart pumping after some catastrophic event, the cessation of the heartbeat was taken as the point of death. When it comes to vampires, the end can be much more spectacular.

According to some films and TV programmes, staking a vampire through the heart causes it to explode into a cloud of dust or instantly disintegrate into a pile of ash. It is very different to what would happen if you were to stake a living person in a vampire costume. It would certainly distinguish between the undead and the living, but such revelations would come too late as far as the law is concerned. It is also quite difficult to do.

Assuming vampires have much the same anatomy as humans, the heart is well protected from damage. The rib cage and breast plate cover the heart and form a strong barrier to be broken through. Considerable force would be needed to drive the stake through the breastbone, but once that has been achieved, there should be more than enough momentum for the stake to continue its journey through the relatively soft tissues underneath.

One blow is recommended. In some parts of the world, especially Russia where they are dealing with the *oupir*, a particularly vicious variety of vampire, failing to destroy the undead with the first blow will only revive it. A half-hearted first attempt may not make it deep enough into the body, but will still cause considerable pain. Not only is the vampire still undead, rather than dead, they are probably quite annoyed. The stakee is, at best, likely to resist further blows and, at worst, be prompted to attack the staker. Many vampire-hunting kits are likely to include a weighty mallet, along with pre-sharpened stakes.

The sharpened point of the stake concentrates the force applied by the mallet to a very small area. Broken broom handles or snapped chair legs are only for dire emergencies when there is no time to refine the broken end to a fine point. The wood the furniture is made from also may not be up to the task. Wood, though it varies enormously between different species of tree, is relatively soft and the point is likely to be blunted as it is forced against the skin and breastbone of the vampire victim. This may be why some woods are preferred for the making of stakes. Some traditions state that using the wrong sort of wood will cause the vampire to come back to terrorise everyone once again.

The mahogany and teak weapons favoured by Blade may have been chosen because of the particular properties of these woods. Mahogany has an even and fine grain that is relatively free of voids, making it very durable. Teak is also tightly grained. This gives the wood strength, and with fewer flaws there are fewer weak points in the wood where a crack could start as you pound the stake into the vampire's chest. However, both these woods grow in the tropics and are not so readily available in Europe, where most of the vampire staking seems to have taken place.

Oak and ash trees are the obvious European choices, as there are plenty of them and both are highly prized for their strength and durability. Ash was the wood of choice for vampire hunters in northern Europe. The wood is generally preferred for things like tool handles that need to be hard wearing and withstand sudden impacts, so it would seem to be a sensible choice for a stake. Ash is preferred because it has similar mechanical properties to oak, but is fast-growing and can be cultivated.[§]

Aspen was favoured by vampire hunters in the Middle Ages. The wood is very easy to work with and has a very uniform texture, meaning it doesn't split easily. In Hungary, the stake must be made of mountain ash, also known as rowan. Rowan is a dense wood that is often used for tool handles and walking sticks, and the trees and shrubs also have a long association with European mythology and folklore. Rowan trees were believed to

---

[§]According to Pliny the Elder in his *Natural History*, all evil things feared ash wood.

have the ability to ward off witches and offered protection against all manner of evil, as well as killing vampires.

Other woods may also have been favoured because of their supposed magical properties rather than, or in addition to, practical physical considerations. Juniper berries have been used to flavour gin, and other parts of the tree have been used in herbal remedies, the smoke from burning the wood has been used in Hogmanay ceremonies, to cleanse and protect homes. The wood has also been used to make stakes to kill vampires.

In parts of Russia, it must be yew wood that is used to destroy the *oupir*. Yew's effectiveness may not just be down to its strength or availability, yew is also very toxic,[¶] a property that has been known about for centuries, and Russian vampire hunters may have been making doubly sure of the complete destruction of their vampires.[||] Almost all parts of the yew contain toxic compounds, the only exception being the red flesh of the berries. And there are several toxic components, but they all fit into the category of taxine alkaloids. These compounds affect the heart cells by changing the way calcium and sodium can flow in and out of cells. The effects on sodium movement affects the electrical signals sent along nerves, which keep the heart beating. The movement of calcium in and out of heart cells is what makes them contract and relax. The alkaloids are easily absorbed from the digestive system, and would take effect even more rapidly if introduced directly into the blood stream. A lethal dose varies by yew species and season but is in the region of 3 mg per kg of body weight. Death is caused by cardiac arrest or respiratory failure.

As the heart seems to be the target organ in defeating vampires, using the toxicity of yew to chemically change the heartbeat makes sense. How much extra damage can be done chemically when there is also a wooden stake puncturing the organ is a moot point. It is certainly a belt-and-braces approach to destroying vampires.

Regardless of the local preference, all woods could be strengthened by heat treatment. The sharpened points in

---

[¶]The Latin name for the whole genus of yew trees is *Taxus*, and is the origin of the word toxic.

[||]Yew swords were also used to kill, or at least contain, the Irish vampire Abhartach, whom we met in Chapter 1.

particular would benefit from burning the tips, to harden them. Holding the wood in a flame will encourage water and other volatile molecules to evaporate, the remaining wood fibres pack closer together making the wood more dense and stronger. Heating too much, however, can cause burning. As the carbon burns, it breaks up compounds, making the wood more brittle and likely to snap or shatter on impact.

It is understandable that some in the past chose to use metal stakes to pin their vampires in place, as shown by bodies that have been excavated, and were found to have a sizeable metal pin still in place. Similar measures have also been taken with fictional vampires. For example, in *Dracula*, the vampire Lucy Westenra was staked with wood, but the more threatening and dangerous Count was destroyed by stabbing his heart with a metal blade.** Metal has the advantages of being stronger than wood, as well as not rotting away over time, which may result in the release of your vampire at a later date. One thirteenth-century grave unearthed in Bulgaria, near the Greek border, was found to contain skeletal remains and an iron rod weighing over two pounds, which had been driven through the shoulder. Around 100 similarly staked burials have been found in Bulgaria. The body of a 40 to 50-year-old man found near the border had also had one leg removed below the knee, and the leg placed next to the body. Dismemberment was yet another way to destroy vampires.

Cutting off the Bulgarian vampire's leg was probably intended to stop him wandering around. But, if he was anything like the Chinese vampire-like creature, the *jiangshi*, he could have hopped.†† Perhaps that's why they took the further measure of staking him down. Mutilations to the body are not uncommon for preventing the undead from rising, as we saw in Chapter 11. But some mutilations are better described as dismemberments, and are clearly intended to kill the undead.

**Although he disintegrated into dust very quickly after the blow, which may mean his body was destroyed, or he transformed himself into a mist just in time and could still be out there.

††You can see for yourself in *Mr Vampire*, a Hong Kong film franchise featuring sinister hopping revenants.

12.6 OFF WITH HIS HEAD!

When it came to putting a stop to vampires once and for all, certain parts of the body came in for particular attention. Most slayers in the past either cut off the head or cut out the heart. Archaeological excavations have uncovered graves where bodies have been found with skulls facing the wrong way, or sometimes the detached head has been placed elsewhere in the coffin or grave site, such as at the feet or by the side of the body. If decapitation could kill the living, why not the dead as well (see Figure 12.1).

Figure 12.1 An anti-vampire burial in Poland. Reproduced from https://commons.wikimedia.org/w/index.php?curid=68490088 under the terms of the CC BY-SA 4.0 license https://creativecommons.org/licenses/by-sa/4.0/deed.en. Reproduced from R. Biskupski – Kotowicz P. (2011), "Wampir" z ulicy Zamkowej 20 w Sanoku, Rocznik Sanocki, t. X, Sanok, s 33–65.

The head is obviously a vital part of the body. People can survive the amputation of limbs, but not the head. Even with the benefits of modern medicine and the advantages of a well-prepared, sterile operating theatre and a dedicated team of medical experts, reattaching a head to a body after it has been separated is unlikely to succeed. A head removed from an already dead body in a graveyard with a sharpened spade has no chance.

Removing the head from the living will result in massive blood loss, and the lack of the all-important oxygen it carries to the brain will lead to brain death in a very short space of time. Even if medical teams could get to work within seconds of the event, critical blood vessels might be repaired, but the damage to the spinal chord would present a very difficult, possibly insurmountable, problem. Nerve cells do not repair or regrow very easily. People who have suffered broken necks may survive the injury, but are often left permanently disabled. Some treatments, if administered quickly, can restore some communication between the brain and the rest of the body beyond the break, but complete restoration after serious injury is beyond the ability of current techniques.

Experiments conducted on monkeys in the 1970s by the American surgeon Robert White showed that surgical techniques were at least good enough for the monkeys to survive a head transplant. He transplanted the heads of rhesus monkeys, and though they survived the operation, they were left as quadriplegics. In 2015, the Italian surgeon Sergio Canavero claimed he had the knowledge to conduct a human head transplant, something he sees as a body transplant. He predicted that he would carry out the first human head transplant in 2017 and that his patient, with physiotherapy, would be up and walking a year later. At the time of writing, the procedure has not been attempted.

Cutting off a head is also hard work, as gruesome executions carried out in the past have shown. Considerable force and a sharp blade is needed to cleave through the strong muscles, tough cartilage and solid bone found in the neck. It is perhaps why some vampire hunters in the past have taken the seemingly easier option of removing another vital organ, the heart.

The heart is also well protected, by the rib cage and breastbone. Evidence from a grave found in Connecticut (see Chapter 8) shows that at least one vampire slayer appears to have accessed the heart through the rib cage, working round the bigger and stronger breast bone. The butcher in the case of the Mykonos vampire elected to use a much softer route – through the abdomen and up (see Chapter 11). This involves less physical strength, but it can be disorientating, and it takes a lot longer, rooting around in the innards.

The practical difficulties and general unpleasantness of removing the heart from a corpse were clearly seen as worth the effort. The heart, as we have seen, is central to our ideas of life and death. Today, hearts can be transplanted between individuals very successfully, but this medical milestone was not reached until the middle of the twentieth century, and is still a complex and risky procedure that is only undertaken in life-threatening situations. Before the advent of skilled stitching techniques in the late nineteenth century, and antibiotics in the early twentieth century, surgery within the main torso of the body was more or less off-limits. Caesarean sections were carried out, but only in the most desperate circumstances and, more often than not, they were fatal for the mother. Lithotomies (removing bladder stones), were carried out with surprising efficiency and success. But surgery on the heart was not even contemplated. Removing the heart from a supposed vampire would be expected to be sufficient to destroy it and end its undead existence. However, it is evident from numerous cases that further steps were often taken once the heart had been extracted. To make absolutely certain and eliminate all threat of the vampire, slayers would burn the organ to ashes.

This extra step may have been necessary for other reasons, such as providing the ashes that could cure those who had been attacked by the vampire. It must have been an important part of the process of ending a vampire epidemic because burning a decayed heart would not be easy. What is even more surprising is that in the most extreme cases, slayers didn't stop at burning just one organ, the whole body had to be consumed by flames to ensure there was no return of the undead.

Cremation would seem a neat and tidy solution to any vampire problem. However, it is actually quite difficult to burn a human body to ashes. It is understandable why many vampire hunters have tried other methods before resorting to building a huge pyre and trying to destroy a whole corpse. But then again, vampires might not be human. Could they be a more combustable human-like species?

12.7 BURN, BABY, BURN

Without detailed knowledge of vampire physiology, we are going to have to assume that they are of similar composition to humans. Humans are mostly water, between two thirds and three quarters of our mass, and water does not burn. The coffin and clothes will burn away relatively quickly but won't provide enough energy to set the body alight, because a lot of energy is needed to drive off the water.

Once the water has been removed by the heat of flames, a body will burn more readily. This is because another large proportion of the human body is made up of carbon-based compounds, and these do burn. Fat, in particular, can provide fuel for the flames, but these fatty compounds are found within the body and a certain amount of destruction has to occur before the skin will burn away or split and release them.

Setting a human body on fire is difficult enough, destroying it completely is even harder. There are many parts of the body that are inorganic and do not burn, because they have already been oxidised, or vaporise like water, such as the components of bone. It is these unburnable inorganic components that form the ashes. Professional crematoria use high temperatures, between 800 °C and 1800 °C, and a steady supply of oxygen to reduce human bodies to just a few kilograms of ash. In a modern crematoria oven, this whole process can be done in less than an hour. The process will still work at lower temperatures, as long as there is a good supply of oxygen. It will just be a lot slower.

Vampires, however, would seem to be considerably more combustible than the average human. In some films, all that is needed is a stray spark for a vampire to be transformed into a flaming torch. Sometimes, a ray of sunshine casually falls on a member of the undead and within seconds they are aflame, at least on screen. Sunlight's association with vampires being set ablaze was down to cinematic special effects (see Chapter 5). Conventional sparks could set clothes on fire and nice long opera capes or funeral shrouds might provide a little extra material, and therefore fuel, for the flames, but the overall contribution would be negligible. But what if there are slight differences between vampire and human cremation that could make the process fractionally easier?

Vampires have usually spent a period of time in their grave before they are dug up and thrown on a bonfire. During their time underground, it is likely they would have gone through a few changes. If the burial site is very dry, the body may have dessicated somewhat to form a gaunt, wizened vampire that contains much less troublesome water that makes fresh bodies so difficult to ignite. Some particularly dry members of the undead may make better kindling, and there is an urban legend that Egyptian mummies were burnt as fuel in the engines of the first trains that crossed Egypt.[‡‡] Even if the story were true, these mummies had been dried out for over three thousand years. Vampires in Europe were dug up much sooner after they had been interred and would have much more damp by comparison.

At the other end of the vampire spectrum from the dry, wizened variety, were the bloated undead that were so characteristic of the eighteenth century. The gases that caused the bloating would be predominantly methane, and that will burn extremely well, but only in the short term. The methane, even if there is a lot of it, will be rapidly consumed in the flames. In fact, it could be so rapid that there might be an explosion if a very bloated vampire were thrown on the fire without being staked to release a lot of the gas first. This would undoubtedly be alarming and unpleasant. The methane would be burned far too quickly and at far too low a temperature to burn the rest of the body, but the explosion might help break it up into smaller chunks that would burn better.

The staking process that could release the built-up gas would also reveal the copious amount of fluid these vampires contained. Though this fluid was, at the time, attributed to the blood sucked form the vampire's victims, it was simply the fluid normally found within the cells of the body released by decomposition processes to slosh about freely. Staking the body before burning might have provided a means of draining some of this fluid before the body was placed on the pyre. It might

[‡‡]A story that probably originated in Mark Twain's piece 'The Innocents Abroad'.

have fractionally reduce the time, effort and fuel needed to burn what was left.

However, there could be an alternative explanation for how all this liquid could in fact fuel the flames. As discussed in previous chapters, if vampires are biologically different to humans, they perhaps have different metabolic processes to support their vampiric lifestyle. It would make sense if they could use anaerobic processes, such as fermentation, to release energy from their food with the limited oxygen available in their coffins (see Chapter 6). One product of fermentation is alcohol that, if it reaches significant concentrations in the vampire's body, would help the combustion process considerably.[§§] But, if this is a possibility, it appears to only apply to certain varieties of vampire. There are plenty of examples from fiction and folklore where vampire bodies proved particularly resistant to the actions of the flames and extreme measures have to be taken.

Varney the Vampire, from Rymer's novel, destroyed himself with fire, but trekked to a volcano to ensure he was completely destroyed by the fiercely hot lava within. Few vampires are so obliging, and not many are likely to be enticed to a nearby volcano or into a crematoria oven to be dispatched. In the past, most vampire burnings were improvised at or near the gravesite. Unsurprisingly, several slayers commented on vampires being 'suspiciously' difficult to reduce to ashes and required 'an enormous amount of wood' or cutting the body into 'nothing but tiny little pieces' before they would combust. As we have seen, this was not suspicious at all.

A pyre built of wood was the most likely scenario, perhaps with oil or pitch added to help the wood burn. A lot of wood and a lot of time will be needed. Flames need fuel and oxygen mixed together intimately, and bodies are dense. This is why some slayers resorted to chopping up the vampire body into little pieces. It increases the flow of oxygen around the individual pieces, but these pieces would still have been reluctantly reduced in the flames because of the water they contained.

After a process that would have taken several hours, the ashes of the vampire were carefully collected and either used as

[§§]Around 12% concentration is the limit for a lot of organisms. Higher alcohol concentrations tend to denature the enzymes that carry out the process.

medicine to treat their victims or scattered, preferably in water. The Bulgarian vampire slayers probably had the right idea about bottling their vampire's first. Sure, it was a lot of effort to get them inside the bottle, but with a better fuel-to-surface-area ratio, it made their final destruction a lot easier.

Cremation is the most certain means of destroying a vampire but the difficulties that go along with it means it has always been seen as a method of last resort. The lessons from this chapter would appear to be prevention is better than cure and, if you are in any doubt about a local vampire problem, leave it to the experts.

Epilogue

So, after all that, what can we conclude about vampires? Well, stories of blood-sucking creatures have certainly been around for a long time. The names they have been given, the descriptions of their appearance and abilities may have mutated over time and geographical location, but there is a common theme of undead beings causing illness and upset among humans. It seems reasonable that there must be some root cause or explanation for all the stories and strange goings on that have been reported so often. What that cause might be is more difficult to ascertain.

Vampires and vampiric traits have been investigated and examined for centuries. Vampirologists past and present have used their knowledge and experience to try and explain these strange phenomena and have offered a whole host of theories, some more credible than others. Vampires have been identified as demons, the restless undead, disease carriers and even a separate species. After searching through lots of different theories and hunting for scientific facts to back them up, I can only say that there may be as many explanations as there are types of vampire. That doesn't make them any less valid, but it certainly makes the journey through them all much more interesting.

Vampirology: The Science of Horror's Most Famous Fiend
By Kathryn Harkup

Published by the Royal Society of Chemistry, www.rsc.org

Some of the characteristics of these creatures can seem too fantastical to be real but, as we have seen, there is often more than a grain of truth behind them. Though humans may not be capable of the feats regularly attributed to vampires, there are animals out there that can live on blood, change their shape, survive in the inhospitable environment of a closed coffin, and defy the normal pattern of ageing. Vampires, or something like them, may be theoretically possible.

Whether you believe vampires are real, or ever have been, is up to you and your interpretation of the events and science described here. And don't just take my word for it. There are many other books and science papers that cover all the aspects discussed here in more detail or from a different perspective. Hopefully, this book has whetted your appetite to find out more, or introduced you to something you had not known before. Whether you have read this book from the comfort of your coffin or skimmed through a few paragraphs in an effort to send you to sleep, at the very least I hope you have enjoyed this short diversion down a strange, dark and twisting path.

Vampires may wish humans harm, but they have inadvertently done us a lot of good. Regardless of whether they could or could not be real, the study of vampires and their characteristics has been of benefit, from new medical treatments to predicting the spread of disease and delaying the signs of ageing. The blood-sucking undead have been a source of huge creativity and entertainment in books, comics, on stage and in films. If nothing else, vampires are interesting in themselves and in the weird and wonderful scientific, historical and cultural paths they can lead you down.

Vampires and vampirology have come a long way. Two centuries of attention from mass media have transformed vampires from decaying corpses to glamorous stars of stage and screen. If there was one moment when vampires went mainstream it was perhaps Hamilton Deane's stage adaptation of *Dracula* in the early twentieth century. It brought the suave and sophisticated vampire to the masses. Deane himself took on the role of the most famous vampirologist of all time, Professor Van Helsing. Who better to end a book on vampirology? At the end of every performance, Deane would appear in front of the curtain and address the audience directly:

'Just a moment, Ladies and Gentlemen! Just a word before you go! We hope the memories of Dracula won't give you bad dreams, so just a word of reassurance. When you get home tonight, and the lights have been turned out, and you are afraid to look behind the curtains, and you dread to see a face appear at the window . . . why just pull yourself together and remember that after all – THERE ARE SUCH THINGS!'

Bibliography

S. Adee, Inside the Clinic Offering Young Blood to Cure Ageing, *New Sci.*, 2017, **235**(3145), 39–43.

C. Ainsworth, Fantastic Beasts, *New Sci.*, 2019, **3251**, 34.

R. S. Ajioka, J. D. Phillips and J. P. Kushner, Biosynthesis of Heme in Mammals, *Biochim. Biophys. Acta*, 2006, **1763**, 72.

M. Allaby, Bats Out of Hell, *New Sci.*, 1998, **2163**, 40.

Anonymous, *Animal Magnetism: Its History to the Present Time. With a Brief Account of the Life of Mesmer*, ed. A. Surgeon and G. B. Dyer, London, 1841.

A. L. Andrew, B. W. Perry, D. C. Card, D. R. Schield and R. P. Ruggiero, *et al.*, Growth and Stress Response Mechanisms Underlying Post-Feeding Regenerative Organ Growth in the Burmese Python, *BMC Genomics*, 2017, **18**, 338.

J.-B. De Boyers Argens (Marquis d'), *The Jewish Spy: Being a Philosophical, Historical and Critical Correspondence: By Letters Which Lately Pass'd Between Certain Jews in Turkey, Italy, France, & c.*, D. Browne and R. Hett, London, 1740, vol. 4.

K. Arney, *Herding Hemingway's Cats: Understanding How Our Genes Work*, Bloomsbury, London, 2017.

P. Barber, *Vampires, Burial, and Death: Folklore and Reality*, Yale University Press, New Haven and London, 1988.

S. Baring-Gould, *The Book of Were-Wolves: Being an Account of a Terrible Superstition*, Smith and Elder, London, 1865.

Vampirology: The Science of Horror's Most Famous Fiend
By Kathryn Harkup

Published by the Royal Society of Chemistry, www.rsc.org

B. Belford, *Bram Stoker: A Biography of the Author of Dracula*, Weidenfeld and Nicolson, London, 1996.

M. E. Bell, *Food for the Dead: On the Trail of New England's Vampires*, Carroll & Graf Publishers, New York, 2001.

A. L. Benabid, T. Costecalde and A. Eliseyev, *et al.*, An Exoskeleton Controlled by an Epidural Wireless Brain-Machine Interface in a Tetraplegic Patient: A Proof of Concept Demonstration, *Lancet*, 2019, **18**(12), 1112.

M. A. Blajchman, Substitutes and Alternatives to Platelet Transfusions in Thrombocytopenic Patients, *J. Thromb. Haemostasis*, 2003, **1**, 1637.

E. Block, *Garlic and Other Alliums: The Lore and the Science*, Royal Society of Chemistry, Cambridge, 2010.

J. Blundell, *Some Remarks on the Operation of Transfusion*, Thomas Tegg, London, 1828.

P. M. Boffey, Rare Disease Proposed as Cause for 'Vampires', *New York Times*, 1985, May 31, Section A, 15.

J. Bondeson, *Buried Alive: The Terrifying History of Our Most Primal Fear*, W.W. Norton and Company, New York, 2001.

R. Bonnet, Photosensitizers of the Porphyrin and Phthalocyanine Series for Photodynamic Therapy, *Chem. Soc. Rev.*, 1995, 19.

A. Brown, *Inside the Wicker Man: How Not to Make a Cult Classic*, Sidgwick and Jackson Ltd., London, 2000.

R. Buffenstein, The Naked Mole-Rat: A New Long-Lived Model for Human Ageing, *J. Gerontol.*, 2005, **60A**(11), 1369.

M. Bunson, *The Vampire Encyclopedia*, Gramercy Books, New York, 1993.

A. Calmet, *Dissertations Upon the Apparitions of Angels, Daemons, and Ghosts, and Concerning the Vampires of Hungary, Bohemia, Moravia, and Silesia*, M. Cooper, London, 1759.

N. Carey, *The Epigenetics Revolution How Modern Biology if Rewriting Our Understanding of Genetics, Disease and Inheritance*, Icon Books Ltd., London, 2011.

C. Cassidy and P. Doherty, *And Then You're Dead: A Scientific Exploration of the World's Most Interesting Ways to Die*, Allen & Unwin, London, 2017.

M. Collins Jenkins, *Vampire Forensics*, National Geographic, Washington D.C., 2010.

L. Da Costa Santos, L. Resende Lucinda, A. da Costa Santos and L. Diniz Silva, Medical Explanations for the Myth of Vampirism, *Rev. Med. Minas. Gerais*, 2013, **23**(4), 510.

D. J. Cripps, Porphyria: Genetic and Acquired, *IARC Sci. Publ.*, 1986, 77, 549.

B. Curran, *Vampires: A Field Guide to the Creatures That Stalk the Night*, New Page Books, Franklin Lakes, 2005.

R. Danovaro, A. Dell'Anno, A. Pusceddu, C. Gambi, A. Heiner and R. M. Kristensen, The First Metazoa Living in Permanently Anoxic Conditions, *BMC Biol.*, 2010, **8**, 30.

M. L. DeBard (Task Force chair), White Paper Report on Excited Delirium Syndrome, *ACEP Excited Delirium Task Force*, 2009.

H. Debruyne, M. Portzky, K. Peremons and K. Audenaert, Cotard's Syndrome: A Review, *Brain, J. Psychiatry*, 2010, **2**(1), 67.

B. J. Edelman, J. Meng, D. Suma, C. Zurn, E. Nagarajan, B. S. Baxter, C. C. Cline and B. He, Noninvasive Neuroimaging Enhances Continuous Meural Tracking for Robotic Device Control, *Robot*, 2019, **4**(31), eaaw6844.

J. Emsley, *The Elements of Murder*, Oxford University Press, Oxford, 2005.

N. M. Foley, G. M. Hughes, Z. Huang and Clarke, *et al.*, Growing Old, Yet Staying Young: The Role of Telomeres in Bats' Exceptional Longevity, *Sci. Adv.*, 2018, **4**(2), eaao0926.

S. L. Forbes, B. H. Stuart and B. B. Dent, The Effect of Burial Environment on Adipocere Formation, *Forensic Sci. Int.*, 2004, **154**, 24.

R. R. Florescu and R. T. McNally, *Dracula, Prince of Many Faces: His Life and His Times*, Back Bay Books, Boston, 1989.

S. L. Forbes, B. B. Dent and B. H. Stuart, The Effect of Soil Type on Adipocere Formation, *Forensic Sci. Int.*, 2005, **154**, 35.

C. Frayling, *Vampyres: Lord Byron to Count Dracula*, Faber and Faber Limited, London, 1978.

R. George, *Nine Pints*, Portobello Books, London, 2018.

E. Gerard, *The Land Beyond the Forest: Facts, Figures and Fancies from Transylvania*, Harper and Brothers, New York, 1888, **2018**.

J. Gómez-Alonso, Rabies: A Possible Explanation for the Vampire Legend, *Neurology*, 1998, **51**(3), 856.

L. M. Gonçalves and A. Tosini, Sudden Onset of Cotard's Syndrome as a Clinical Sign of Brain Tumour, *Arch. Clin. Psychiatry*, 2016, **43**(2), 35.

A. Gregory, *Nodding Off: The Science of Sleep from Cradle to Grave*, Bloomsbury Publishing, London, 2018.

N. Groom, *The Vampire: A New History*, Yale University Press, New Haven, 2018.

R. Halder, D. Gupta, S. Chitranshi, M. Kumar-Singh and S. Sachan, Artificial Blood: A Futuristic Dimension of Modern Day Transfusion Sciences, *Cardiovasc. Hematol. Agents Med. Chem.*, 2019, **17**, 11.

C. Hamlin, *Cholera: The Biography*, Oxford University press, Oxford, 2009.

J. S. Hampi and W. S. Hampi, Pellagra and the Origin of a Myth: Evidence from European Literature and Folklore, *J. R. Soc. Med.*, 1997, **90**, 636–639.

W. R. Hansen and K. Autumn, Evidence for Self-Cleaning in Gecko Setae, *Proc. Natl. Acad. Sci. U. S. A.*, 1997, **102**(2), 385.

E. W. Hawkes, E. V. Eason, D. L. Christensen and M. R. Cutkosky, Human Climbing With Efficiently Scaled Gecko-Inspired Dry Adhesives, *J. R. Soc., Interface*, 2015, **12**, 20140675.

D. Hayden, *Pox: Genius, Madness, and the Mysteries of Syphilis*, Basic Books, New York, 2003.

J. Hayman and M. Oxenham, *Human Body Decomposition*, Elsevier, London, 2016.

A. Helldén, Death Delusion, *BMJ*, 2007, **335**, 1305.

J. F. Y. Hoh, G. H. Rossmanith, L. J. Kwan and A. M. Hamilton, Adrenaline Increases the Rate of Cycling of Crossbridges in Rat Cardiac Muscle as Measured by Pseudo-Random Binary Noise-Modulated Perturbation Analysis, *Circ. Res.*, 1988, **62**(3), 452.

J. B. Holcomb, Optimal Use of Blood Products in Severely Injured Trauma Patients, *Hematol. Am. Soc. Hematol. Educ. Prog.*, 2010, 465.

D. W. Huestis, Alexander Bogdanov: The Forgotten Pioneer of Blood Transfusion, *Transfus. Med. Rev.*, 2007, **21**(4), 337.

L. Illis, On Porphyria and the Ætology of Werwolves, *Proc. R. Soc. Med.*, 1964, **57**, 23.

M. R. James, *Count Magnus and Other Stories*, Penguin Books, London, 2005.

F. Kaplan, 'The Mesmeric Mania': The Early Victorians and Animal Magnetism, *J. Hist. Ideas*, 1975, **35**(4), 691.

A. King, Hanging Tough, *New Sci.*, 2016, **3060**, 34.

C. Kyba, A. Muhar and T. Posch, How Bright is Moonlight? *Astron. Geophys.*, 2017, **58**(1), 31.

P. Lamont, Reflexivity, the Role of History, and the Case of Mesmerism in Early Victorian Britain, *Hist. Psychol.*, 2010, **13**(4), 393.

N. Lane, Born to the Purple: the Story of Porphyria, *In Focus*, 2002.

N. Lane, Therapy: New Light on Medicine, *Scientific American*, 2007.

S. Le Fanu, *Carmilla*, Echo Library, Middlesex, 2006 (Year of original publication: 1872).

G. T. Liberatore, A. Samson, C. Bladin, W.-D. Schleuning and R. L. Medcalf, Vampire Bat Salivary Plasminogen Activator (Desmoteplase): A Unique Fibrinolytic Enzyme That Does Not Promote Neurodegeneration, *Stroke*, 2003, 537.

The Secret Science of Superheroes, ed. M. Lorch and A. Miah, Royal Society of Chemistry, London, 2017.

K. Lougheed, *Catching Breath: The Making and Unmaking of Tuberculosis*, Bloomsbury, London, 2017.

M. Lyman, *The Remarkable Life of the Skin: An Intimate Journey Across Our Surface*, Bantam Press, London, 2019.

F. MacCarthy, *Byron: Life and Legend*, Faber and Faber Ltd., England, 2003.

F. Marryat, *The Blood of the Vampire*, Bernhard Tauchnitz, Leipzig, 1897.

R. Matheson, *I Am Legend*, Orion Publishing Group, London, 2010 (Year of original publication: 1954).

C. R. Maturin, *Melmoth, The Wanderer: A Tale*, Archibald Constable and Company, and Hurst, Robinson and Co., London, 1821.

H. Mayo, *Letters on the Truths Contained in Popular Superstitions*, Messrs. Blackwood, Edinburgh, 1849.

W. H. McNeill, *Plagues and Peoples*, Anchor Press/Doubleday, New York, 1976.

S. Meyer, *Twilight*, Atom Books, London, 2009.

E. Miller, *A Dracula Handbook*, Xlibris Corporation, 2005.

R. Montillo, *The Lady and Her Monsters: A Tale of Dissections, Real-Life Dr. Frankensteins, and the Creation of Mary Shelley's Masterpiece*, HarperCollins Publishers, New York, 2013.

T. Moore, *The Life, Letters and Journals of Lord Byron*, John Murray, London, 1860.

M. D. Norman, J. Finn and T. Tregenza, Dynamic Mimicry in an Indo-Malayan Octopus, *Proc. R. Soc. London, Ser. B*, 2001, **268**, 1755.

E. Nuzzolese and M. Borrini, Forensic Approach to an Archaeological Casework of 'Vampire' Skeletal Remains in Venice: Odontological and Anthropological Prospectus, *J. Forensic Sci.*, 2010, **55**(6), 1634.

M. C. O'Neill, B. R. Umberger, N. B. Hulowka, S. G. Larson and P. J. Reiser, Chimpanzee Super Strength and Human Skeletal Muscles Evolution, *Proc. Natl. Acad. Sci. U. S. A.*, 2017, **28**, 7343.

J. Oullette, *The Physics of the Buffyverse*, Penguin Books Ltd., London, 2006.

J. W. Polidori, *The Vampyre*, Travelman Publishing, London, 1999 (Year of original publication: 1819).

The Diary of Dr. John William Polidori, 1816, ed. J. W. Polidori and W. M. Rossetti, Elkin Matthews, London, 1911.

J. Prahlow, *Forensic Pathology for Police, Death Investigators, Attorneys, and Forensic Scientists*, Springer, London, 2010.

C. Quigley, *The Corpse: A History*, McFarland & Company, Inc., North Carolina, 2015.

K. Ramsland, *The Science of Vampires*, Berkley Boulevard Books, New York, 2002.

D. Reich, R. Green and M. Kircher, *et al.*, Genetic History of an Archaic Hominin Group from Denisova Cave in Siberia, *Nature*, 2010, **468**, 1053.

J. Reiger, Dr Polidori and the Genesis of Frankenstein, *Studies in English Literature 1500–1900*, 1963, **3**(4), 461.

A. Rice, *Interview With the Vampire*, Future Books, London, 1994.

A. Rice, *The Vampire Lestat*, Future Books, London, 1995.

M. Roach, *Stiff: The Curious Lives of Human Cadavers*, Penguin Books, London, 2004.

T. R. Robinson, *Genetics for Dummies*, John Wiley and Sons, Inc., Hoboken, 2nd edn, 2010.

R. Rothbaum and J. S. McGee, Aquagenic Urticaria: Diagnostic and Management Challenges, *J. Asthma Allergy*, 2016, **9**, 209.

P. Rothemund, Hemin and Chlorophyll - The Two Most Important Pigments for Life on Earth, *Ohio J. Sci.*, 1956, **LVI**(4), 193.

J. M. Rymer, *Varney, the Vampire or, The Feast of Blood*, Wordsworth Editions Limited, Hertfordshire, 2010 (Year of original publication: 1845).

P. Saukko and B. Knight, *Knight's Forensic Pathology*, Hodder Arnold Ltd., London, 3rd edn, 2004.

G. Schierhout and I. Roberts, Fluid Resuscitation with Colloid or Crystalloid Solution in Critically Ill Patients: A Systematic Review of Randomised Trials, *BMJ*, 1998, **316**, 961.

D. J. Skal, *Vampires: Encounters With the Dead*, Black Dog & Leventhal, New York, 2006.

P. S. Sledzik and N. Bellantoni, Bioarcheological and Biocultural Evidence for the New England Vampire Folk Belief, *Am. J. Phys. Anthropol.*, 1994, **94**, 269.

M. Smith, Osiris and the Deceased, *UCLA Encyclopedia of Egyptology*, 2008, **1**(1), 2.

M. Stevens and S. Merilaita, *Animal Camouflage: Mechanisms and Function*, Cambridge University Press, Cambridge, 2011.

B. Stoker, *Dracula*, Wordsworth Editions Limited, Hertfordshire, 1993 (Year of original publication: 1897).

W. Strieber, *The Hunger*, Transworld Publishers, London, 1982.

C. Stringer, *The Origin of Our Species*, Penguin Books, London, 2011.

N. Suckling, *Vampires*, AAPPL Artists' and Photographers' Press Ltd., London, 2006.

A. M. Summers, *The Vampire, His Kith and Kin*, Forgotten Books, 2008 (Year of original publication: 1928).

A. M. Summers, *The Vampire in Europe*, Random House, London, 1996 (Year of original publication: 1928).

M. D. Sztajnkrycer and A. A. Baez, Cocaine, Excited Delirium and Sudden Unexpected Death, *Emerg. Med. Serv.*, 2005, 77.

T. Taylor, The Real Vampire Slayers, *The Independent*, 2007.

D. Teresi, *The Undead*, Vintage Books, New York, 2012.

J. Teyssier, S. V. Saenko, D. van der Marel and M. C. Milinkovitch, Photonic Crystals Cause Active Colour Change in Chameleons, *Nat. Commun.*, 2015, **6**, 6368.

H. Thadani, A. Deacon and T. Peters, Diagnosis and Management of Porphyria, *BMJ*, 2000, **320**, 1657.

D. Thomas, Vampire Bats 'Kill Villagers', *Telegraph*, 2008.

H. Thomson, D. Teresi, S. Cave and C. Williams, Special Issue: Death, *New Sci.*, 2012, **2887**, https://www.newscientist.com/issue/2887/.

H. Thomson, The Way We Die Now, *New Sci.*, 2019, 3257. https://www.newscientist.com/article/mg24432570-300-why-the-line-between-life-and-death-is-now-more-blurred-than-ever/.

N. L. Tilney, *Transplant: From Myth to Reality*, Yale University Press, New Haven and London, 2003.

J. P. De Tournefort, *A Voyage into the Levant: Perform'd by the Command of the late French King. Containing the Ancient and Modern State of the Islands of the Archipelago; as Also of Constantinople, the Coasts of the Black Sea, Armenia, Georgia, the Frontiers of Persia and Asia Minor*, D. Midwinter, London, 1741, vol. 1.

M. Tuttle, *The Secret Lives of Bats: My Adventures With the World's Most Misunderstood Mammals*, Mariner Books, New York, 2015.

C. Valentine, *Past Mortems: Life & Death Behind Mortuary Doors*, Little, Brown Book Group, London, 2017.

T. Vampirologist, *The Case of Petre Toma*, https://vampirologist.blogspot.com/2009/01/case-of-petre-toma.html.

A. De Verteuil, *Scientific Sorties*, The Litho Press, Trinidad, 1993.

S. A. Villeda, K. E. Plambeck and J. Middeldorp, *et al.*, Young Blood Reverses Age-Related Impairments in Cognitive Function and Synaptic Plasticity in Mice, *Nat. Med.*, 2014, **20**(6), 659.

E. Wadge, The Scientific Spirit and the Spiritual Scientist: Moving in the Right Circles, *Vic. Rev.*, 2000, **25**(1), 24.

B. Wasik and M. Murphy, *Rabid: A History of the World's Most Diabolical Virus*, Penguin Books Ltd., London, 2012.

J. C. Whorton, *The Arsenic Century: How Victorian Britain was Poisoned at Home, Work, & Play*, Oxford University Press, Oxford, 2010.

W. Wilkinson, *An Account of the Principalities of Wallachia and Maldovia*, Longman, Hurst, Rees, Orme and Brown, London, 1820.

A. Williams, Death: The Great Bacterial Takeover, *New Sci.*, 2014, **2984**, 10.

K. M. Wilson, The History of the Word 'Vampire', *J. Hist. Ideas*, 1985, **46**(4), 577.

M. S. Wong, W. J. Hawthorne and N. Manolios, Gene Therapy in Diabetics, *Self/Nonself*, 2010, **1**(3), 165.

C. Wynne, Mesmeric Exorcism, Idolatrous Beliefs, and Bloody Rituals: Mesmerism, Catholicism, and Second Sight in Bram Stoker's Fiction, *Vic. Rev.*, 2000, **26**(1), 43.

C. Zimmer, *Soul Made Flesh: How the Secrets of the Brain were Uncovered in Seventeenth-Century England*, Arrow Books, London, 2005.

WEBSITES

- Information about porphyria: https://porphyria.eu/
- Information about rabies: https://www.nhs.uk/conditions/Rabies/

Subject Index

References to figures are given in *italic* type. Notes are indicated by an 'n' following the page number.